Math Girls⁴

Randomized Algorithms

Hiroshi Yuki

Translated by Tony Gonzalez

http://bentobooks.com

MATH GIRLS 4: RANDOMIZED ALGORITHMS
by Hiroshi Yuki

Originally published as *Sūgaku Gāru Rantaku Arugorizumu*
Copyright © 2011 Hiroshi Yuki
Softbank Creative Corp., Tokyo

English translation © 2020 by Tony Gonzalez
Edited by M.D. Hendon
Cover design by Kasia Bytnerowicz

All rights reserved. No portion of this book in excess of fair use considerations may be reproduced or transmitted in any form or by any means without written permission from the copyright holders.

Published 2020 by

Bento Books, Inc.
Austin, Texas 78732

bentobooks.com

ISBN 978-1-939326-45-4 (hardcover)
ISBN 978-1-939326-43-0 (trade paperback)
ISBN 978-1-939326-44-7 (case laminate)

Printed in the United States of America
First edition, March 2020

Math Girls^4:
Randomized Algorithms

Contents

1 A Bet You Cannot Lose   5
     1.1 Throwing Dice   5
         1.1.1 Two Dice   5
     1.2 Coin Tosses   8
         1.2.1 Two Coins   8
         1.2.2 One Coin   10
         1.2.3 Lotteries and Memory   11
     1.3 The Monty Hall Problem   13
         1.3.1 Three Doors   13
         1.3.2 A God's-Eye View   19

2 Step By Step   23
     2.1 At School   23
         2.1.1 Tetra   23
         2.1.2 Lisa   24
         2.1.3 Linear searches   25
         2.1.4 A Walkthrough   28
         2.1.5 Analyzing the Linear Search   32
         2.1.6 Analysis of Linear Search When v is in the List   33
         2.1.7 Analysis of Linear Search When v is Not in the List   35
     2.2 Analysis of Algorithms   37

	2.2.1	Miruka 37

- 2.2.1 Miruka 37
- 2.2.2 Characteristics of Good Algorithms 38
- 2.2.3 Eliminating Cases 39
- 2.2.4 Considering Meaning 41
- 2.2.5 Sentinel Linear Search 44
- 2.2.6 Creating History 47

2.3 At Home 49
- 2.3.1 A Step Forward 49

3 The Solitude of 17,179,869,184 55
- 3.1 Permutations 55
  - 3.1.1 At the Book Store 55
  - 3.1.2 Getting It 56
  - 3.1.3 An Example 57
  - 3.1.4 Regularity 58
  - 3.1.5 Generalization 63
  - 3.1.6 Forging Paths 65
  - 3.1.7 This Guy 66
- 3.2 Combinations 67
  - 3.2.1 In the Library 67
  - 3.2.2 Permutations 68
  - 3.2.3 Combinations 70
  - 3.2.4 Animal 72
  - 3.2.5 The Binomial Theorem 74
- 3.3 Allocations of 2^n 77
  - 3.3.1 Pascal's Triangle 77
  - 3.3.2 Bit Patterns 81
  - 3.3.3 Exponential Explosions 83
- 3.4 The Solitude of Powers 84
  - 3.4.1 On the Road Home 84
  - 3.4.2 At Home 85

4 Uncertain Certainty 87
- 4.1 Certain Certainty 87
  - 4.1.1 What Division Means 87
- 4.2 Uncertain Certainty 92
  - 4.2.1 Same Certainties 92

- 4.2.2 The True Weapon 93
- 4.3 An Experiment in Certainty 95
  - 4.3.1 Interpreter 95
  - 4.3.2 The Dice Game 97
  - 4.3.3 Roulette 98
- 4.4 Destruction of Certainty 99
  - 4.4.1 Defining Probability 99
  - 4.4.2 Kinds of Probability 102
  - 4.4.3 Mathematical Applications 102
  - 4.4.4 Three Perspectives 103
- 4.5 Axiomatic Definition of Certainty 104
  - 4.5.1 Andrey Kolmogorov 104
  - 4.5.2 Sample Spaces and Probability Distributions 105
  - 4.5.3 The Probability Axioms 108
  - 4.5.4 Subsets and Events 109
  - 4.5.5 Probability Axiom P1 110
  - 4.5.6 Not Quite There 114
  - 4.5.7 Even Odds 115
  - 4.5.8 Loaded Dice and Edgy Coins 117
  - 4.5.9 Promises 118
  - 4.5.10 Coughs 119

5 Expected Values 121
- 5.1 Random Variables 121
  - 5.1.1 Mom 121
  - 5.1.2 Tetra 122
  - 5.1.3 Examples of Random Variables 124
  - 5.1.4 Examples of Probability Distributions 127
  - 5.1.5 So Many Words 129
  - 5.1.6 Expected Values 129
  - 5.1.7 Fair Game 134
- 5.2 Linearity 135
  - 5.2.1 Miruka 135
  - 5.2.2 The Expected Value of a Sum is the Sum of Expected Values 135

- 5.3 The Binomial Distribution 140
  - 5.3.1 Speaking of Coins 140
  - 5.3.2 Expected Value of a Binomial Distribution 142
  - 5.3.3 Separation into Sums 144
  - 5.3.4 Indicator Random Variables 146
  - 5.3.5 Some Fun Homework 147
- 5.4 Everything Happening 148
  - 5.4.1 Someday 148
  - 5.4.2 Throwing Dice 149
  - 5.4.3 The Stairway of Happiness 152
  - 5.4.4 Measuring the Staircase 154
  - 5.4.5 Slip of the Tongue 160

6 Indiscernible Future 165

- 6.1 Memory of a Promise 165
  - 6.1.1 On a Riverbank 165
- 6.2 Order 166
  - 6.2.1 Fast Algorithms 166
  - 6.2.2 Of Order At Most n 168
  - 6.2.3 A Quiz 171
  - 6.2.4 Of Order at Most $f(n)$ 173
  - 6.2.5 $\log n$ 177
- 6.3 Search 179
  - 6.3.1 Binary Search 179
  - 6.3.2 An Example 182
  - 6.3.3 Analysis 184
  - 6.3.4 A Promise of Sorts 190
- 6.4 Sorting 190
  - 6.4.1 Bubble Sort 190
  - 6.4.2 An Example 192
  - 6.4.3 Analysis 194
  - 6.4.4 Big-O hierarchy 198
- 6.5 Dynamic Perspectives, Static Perspectives 199
  - 6.5.1 Number of Needed Comparisons 199
  - 6.5.2 Comparison Trees 201

         6.5.3   Evaluating log n!   203
   6.6   Passing On and Learning   206
         6.6.1   Passing On   206
         6.6.2   Learning   207

7 Matrices   211
   7.1   In the Library   211
         7.1.1   Ms. Mizutani   211
         7.1.2   Organic Tetras   212
   7.2   Yuri   213
         7.2.1   Inconsistent   213
         7.2.2   Underdetermined   214
         7.2.3   Regular   216
         7.2.4   A Letter   226
   7.3   Tetra   227
         7.3.1   In the Library   227
         7.3.2   Rows and Columns   227
         7.3.3   Matrix–Vector Products   228
         7.3.4   Systems of Equations and Matrices   230
         7.3.5   Matrix Products   231
         7.3.6   Invertible matrices   232
   7.4   Miruka   236
         7.4.1   Finding Hidden Mysteries   236
         7.4.2   Linear Transformations   241
         7.4.3   Rotations   248
   7.5   Heading Home   250
         7.5.1   Dialogues   250

8 A Random Walk Alone   255
   8.1   At Home   255
         8.1.1   A Rainy Saturday   255
         8.1.2   Teatime   256
         8.1.3   Piano Problem   256
         8.1.4   Example Melody   258
         8.1.5   Solution 1: Hard Work   260
         8.1.6   Solution 2: Smart Work   262
         8.1.7   Generalization   267

- 8.1.8 Apprehension (1) 271
- 8.2 On the Way to School 272
  - 8.2.1 Random Walks 272
- 8.3 Lunch in Our Classroom 273
  - 8.3.1 Practicing with Matrices 273
  - 8.3.2 Apprehension (2) 276
- 8.4 In the Library 278
  - 8.4.1 The Wandering Problem 278
  - 8.4.2 The Meaning of A^2 282
  - 8.4.3 The nth Power of a Matrix 283
  - 8.4.4 Some Prep: Diagonal Matrices 284
  - 8.4.5 More Prep: Matrix and Inverse Matrix Sandwiches 285
  - 8.4.6 To Eigenvalues 286
  - 8.4.7 Eigenvectors 291
  - 8.4.8 Finding A^n 293
- 8.5 At Home 296
  - 8.5.1 Apprehension (3) 296
  - 8.5.2 A Rainy Night 297

9 Strongly, Correctly, Beautifully 299
- 9.1 At Home 299
  - 9.1.1 A Rainy Saturday 299
- 9.2 In the Library 304
  - 9.2.1 The Logical Approach 304
  - 9.2.2 The Satisfiability Problem 305
  - 9.2.3 3-SAT 306
  - 9.2.4 Satisfying 309
  - 9.2.5 Assignment Practice 309
  - 9.2.6 NP-complete problems 310
- 9.3 On the Way Home 312
  - 9.3.1 Oaths and promises 312
  - 9.3.2 The Conference 313
- 9.4 In the Library 314
  - 9.4.1 A Randomized Algorithm for Solving 3-SAT 314
  - 9.4.2 Random Walks 315

- 9.4.3 Quantitative estimations 319
- 9.4.4 Another Random Walk 320
- 9.4.5 Focusing on Rounds 322
- 9.5 At Home 326
  - 9.5.1 Estimating Luck 326
  - 9.5.2 Simplifying the Sum 330
  - 9.5.3 Progress 332
- 9.6 In the Library 332
  - 9.6.1 Independence and Exclusion 332
  - 9.6.2 Precise Estimates 333
  - 9.6.3 Stirling's Approximation 337
- 9.7 On the Way Home 342
  - 9.7.1 Olympics 342
- 9.8 At Home 344
  - 9.8.1 Logic 344

10 Randomized Algorithms 347

- 10.1 At the Restaurant 347
  - 10.1.1 Rain 347
- 10.2 At School 349
  - 10.2.1 Lunchtime 349
  - 10.2.2 The Quicksort Algorithm 349
  - 10.2.3 Array Partitioning with a Pivot: Two Wings 353
  - 10.2.4 Sorting subarrays—Recursion 356
  - 10.2.5 Analysis of Execution Steps 357
  - 10.2.6 By Cases 359
  - 10.2.7 Maximum Execution Steps 363
  - 10.2.8 Average Execution Steps 367
  - 10.2.9 On the Way Home 372
- 10.3 At Home 372
  - 10.3.1 Changing Forms 372
  - 10.3.2 H_n and $\log n$ 378
- 10.4 In the Library 379
  - 10.4.1 Miruka 379
  - 10.4.2 Randomized Quicksort 380
  - 10.4.3 Observing Comparisons 383

- 10.4.4 Linearity of expectation 386
- 10.4.5 The Expected Value of an Indicator Random Variable is its Probability 387
- 10.5 At the Restaurant 390
  - 10.5.1 Other Randomized Algorithms 390
  - 10.5.2 Preparation 390
- 10.6 At the Narabikura Library 391
  - 10.6.1 Iodine 391
  - 10.6.2 Anxiety 392
  - 10.6.3 The Presentation 393
  - 10.6.4 Passing on 394
  - 10.6.5 Oxygen 395
  - 10.6.6 Connecting 396
  - 10.6.7 In the Garden 397
  - 10.6.8 Symbol of a Promise 398

Recommended Reading 407

Index 417

To my readers

This book contains math problems covering a wide range of difficulty. Some will be approachable by elementary school students, while others will challenge even college students.

The characters often use words, diagrams, and computer programs to express their thoughts, but in some places equations tell the tale. If you find yourself faced with math you don't understand, feel free to just browse through it before continuing with the story. Tetra and Yuri will be there to keep you company.

If you have some skill at mathematics, then please follow not only the story, but also the math. Doing so is the best way to fully engage in the tale that is being told.

—Hiroshi Yuki

Prologue

> There are no paths before me.
> Paths form where I tread.
>
> — Kotaro Takamura
> *Michinori*

I want to know the world.
I want to know myself.

    I seek to comprehend the breadth of the world,
    and the depth of my soul.

Even more, I want to be understood.
I want the world—I want *her*—to know me.
Yet even I do not understand.
I do not understand myself.
Do I truly wish to show myself as I am?
I don't even know that.

    The quiet girl with the red hair.
    New encounters in new seasons, bringing new mysteries.

By choosing one I forgo the other.
I must choose one of infinitely many paths.

    The past is set, the future uncertain,
    and we stand on the border between the two.

The future becomes the present, a single moment.
Now is a trail we leave as we tread
our path into an uncertain future.

 Our choices form our course, leading into the future.

I must choose, even without understanding.
I must live, even without comprehension.
Through choice I live, by advancing I create my path.

 There are no paths before me.
 Paths form where I tread.

I do not know if my understanding of the world is true.
I do not know if I truly understand myself.
Yet today too, I shall continue on.

 To know tomorrow.
 To reveal the mysterious.

To share my dreams of a future with you.

Chapter 1

A Bet You Cannot Lose

> My next work was to view the country, and seek a proper place for my habitation, and where to stow my goods to secure them from whatever might happen. Where I was, I yet knew not; whether on the continent or on an island; whether inhabited or not inhabited; whether in danger of wild beasts or not.
>
> DANIEL DEFOE
> *Robinson Crusoe*

1.1 THROWING DICE

1.1.1 Two Dice

"I've got a problem for you!" Yuri shouted.
 "Of course you do," I said. "What is it today?"
 "A good one! Prepare to be humiliated!"

> Alice and Bob are playing a dice game: each throws a single die, and whoever gets the larger number wins. What is the probability that Alice will win?

My cousin Yuri lived in my neighborhood, and since we were little she'd hung out with me so often many thought we were siblings. She would soon be entering her third and final year of junior high. Her chestnut brown hair was pulled back into a ponytail, and she was wearing jeans with a thin sweater.

It was April, a time of gently warm weather. We were in my room, as was frequently the case; recently we had taken to doing math together, trying to trip each other up with problems like this.

"Even odds, right?" I said. "So shouldn't the answer be $\frac{1}{2}$?"

From the gleeful look on Yuri's face, I immediately knew I was wrong.

"Oh, of course," I said. "They could—"

"They could tie!"

"Okay, you got me. So let's do it right. There are 6 ways Alice could throw her die, and 6 ways Bob could throw his. That means there's $6 \times 6 = 36$ possible patterns in all, each with an equal probability of coming up."

$$\text{All possible cases} = 6 \text{ cases for Alice} \times 6 \text{ cases for Bob}$$
$$= 36 \text{ possibilities}$$

Yuri nodded, and I continued.

"Of those 36 possibilities, in 6 of them Alice and Bob will throw the same number, resulting in a tie. That means there's a clear winner in $36 - 6 = 30$ cases, half of which Alice will win and half of which Bob will win."

$$\begin{aligned}
\text{Cases where Alice wins} &= \frac{\text{Cases with a clear winner}}{2} \\
&= \frac{\text{All possible cases (36)} - \text{Cases of a tie (6)}}{2} \\
&= \frac{36 - 6}{2} \\
&= \frac{30}{2} \\
&= 15
\end{aligned}$$

"Yep, yep," Yuri said.

"So the probability of an Alice win would be this."

$$\text{Probability of Alice winning} = \frac{\text{Cases where Alice wins (15)}}{\text{All cases (36)}}$$
$$= \frac{15}{36}$$
$$= \frac{5}{12}$$

"Right-o," Yuri said. "The probability that Alice wins is $\frac{5}{12}$. But finally getting the correct answer still doesn't excuse you from falling into the 'I forgot about ties' trap."

"It happens."

"Remember, you're the one who's always going on about using tables to simplify tricky problems."

		⚀ Bob	⚁	⚂	⚃	⚄	⚅
	⚀	Tie	Bob	Bob	Bob	Bob	Bob
	⚁	Alice	Tie	Bob	Bob	Bob	Bob
Alice	⚂	Alice	Alice	Tie	Bob	Bob	Bob
	⚃	Alice	Alice	Alice	Tie	Bob	Bob
	⚄	Alice	Alice	Alice	Alice	Tie	Bob
	⚅	Alice	Alice	Alice	Alice	Alice	Tie

Alice and Bob's dice game.

"I am indeed," I said, gritting my teeth at having made such a stupid mistake.

"You get the $\frac{15}{36} = \frac{5}{12}$ like this, right?"

$$= \frac{15}{36} = \frac{5}{12}$$

"Uh, yeah. Okay, smarty pants, it's my turn to give a problem."
"Sure, but no picking on me, now."

1.2 Coin Tosses

1.2.1 Two Coins

Tossing 2 coins

Alice tosses one ¥100 coin and one ¥10 coin in secret, and says, "At least one of the coins came up heads." What is the probability that both coins came up heads?

"Easy peasy," Yuri said.
"Yeah?"
"We already know at least one coin came up heads, so we just need to think about the probability that the other one is heads too. So the answer is $\frac{1}{2}$. Done."
"Not done. Not correctly, at least."
"What the—" Yuri said, clearly taken aback. "That's gotta be right."

"Nope."

"It is!"

"Really, it isn't. When you're doing probability problems, you have to step back and look at the big picture."

"Yep, and the picture here is a big, fat $\frac{1}{2}$."

"Seriously, it isn't. Look, let's make a table of all the possibilities."

	¥100 coin	¥10 coin
HH	Heads	Heads
HT	Heads	Tails
TH	Tails	Heads
TT	Tails	Tails

"What's this HH and all?"

"H is for 'heads' and T is for 'tails,' so 'HH' means 'heads–heads,' etcetera."

"Right, gotcha."

"So anyway, if you throw two coins, you have an equal probability of getting one of four patterns: HH, HT, TH, or TT."

"Well yeah, but in this case it can't be TT, since Alice told us that there's at least one heads."

"She did. So we know that the toss ended up as one of HH, HT, or TH."

	¥100 coin	¥10 coin
HH	Heads	Heads
HT	Heads	Tails
TH	Tails	Heads
~~TT~~	~~Tails~~	~~Tails~~

"Oops."

"Each of HH, HT, and TH can occur with equal probability of $\frac{1}{3}$. Among those only HH means that both turned up heads, so the answer is $\frac{1}{3}$."

"Mmm..." Yuri grimaced, sunlight flashing golden off her hair. "Remind me once more what Alice said?"

"She said, 'At least one of the coins came up heads.'"

"Ah, I've got it! It's that 'at least one of' part that's key, isn't it? It could have been the ¥100 coin, and it could have been the ¥10 coin. There's two ways it could happen."

"Yeah, sure. And there's three ways that at least one heads could show up, among which only HH has both of the coins showing heads. In the other two cases—HT and TH—only one is heads."

"Okay, I get it."

"Like I said before, you have to look at the big picture."

1.2.2 One Coin

Yuri stretched out her arms and gave an enormous yawn.

"I'm getting tired of all this theorizing. Let's play a real coin game. Gimme ¥100."

I handed Yuri a coin, which she flicked into the air with a metallic *ting*. It flew straight up, then fell back down, flashing silver. She caught it on the back of her left hand and immediately covered it with her right.

"You're good at that," I said.

"Okay, heads I win, tails you lose."

"Sure. No, wait..."

"Caught on already?"

"Well of course I did. There's no way I can win!"

"Sure you can. If it lands on its side."

I sighed with a smile.

"Okay, okay. Seriously, then. Heads or tails?"

"Tails."

Yuri removed her hand.

"Heads," she said. "I win."

"Uh, that's the back, which is tails. So *I* win."

"Huh? How's that?"

"The side with the date on it is the back. That's how the Japanese mint defines it, anyway."

"You sure about that?" Yuri said, eyeing me suspiciously.

"Go look it up," I said.

Heads   Tails

"I'd rather give you a harder problem. How about this..."

Just then, my mother called from the kitchen.

"Hey, kids! Snack time!"

"Coming!" Yuri shouted back. "Looks like you were saved by the bell."

"Saved. Right."

"Anyway, c'mon!" she said, and dashed for the door.

I slumped along after her.

1.2.3 Lotteries and Memory

When I arrived at the table, I saw a plate of crackers and bowls of hot soup.

"What's this?" I asked, giving a suspicious sniff.

"An experiment, using some new spices I picked up," my mother replied. "Tell me what you think."

"I think it smells weird," I said.

"I think it smells great!" Yuri cooed.

"Such a lovely child," my mother said.

I took a cautious sip and choked. "That's... awful."

"How rude!" my mother said. "Like you're one to talk. Remember that cream soup I let you make the other day? Now *that* had a memorable flavor, to be sure."

"That you *let* me make? You forced me to make it! Besides, you said to tell you what I think."

"All he had to do was stir it," she said to Yuri. "Oh, and remember when he was supposed to watch over the beans at New Year's? He managed to burn even the pot!"

"I was... caught up in a book that day."

"Time to sharpen your housekeeping skills, cuz," Yuri said. "I might need them some day."

"You might... Hang on, now."

"Is he teaching you well, dear?" my mother asked Yuri.

"He is! We were just working on some probability problems," she said with a smile, not missing a beat.

"Oh! Speaking of probability," Mother said, "at the train station today I saw a kiosk selling tickets for the spring lottery."

"I saw that," I said. "It was on a big poster that said 'A grand prize ticket was purchased here,' right?"

I pictured the handwritten sign, underline and all.

Spring Lottery!
A <u>grand prize ticket</u> was purchased here!

"Maybe I'll finally get lucky if I buy a ticket there."

"That's not how it works, Mom," I said. "Just because they sold a winning ticket once, doesn't mean they'll do it again."

"I don't know. Maybe they have a higher probability of having winners."

"Mathematically speaking, that's not the case."

"Might be, if you factor in luck."

"No, Mom, don't let the poster fool you. Lotteries don't have memories, no recollection of where previous winners showed up."

"If you say so," she said, clearly unconvinced.

Yuri pushed aside her empty bowl.

"The poster just said that a winning ticket was purchased at that kiosk, right?" she said.

"It did," I said.

"And there's nothing wrong about that, mathematically speaking."

"Well, no," I said, unsure where Yuri was going with this.

"So you can't really claim that the poster is trying to fool anyone. It's just stating the fact that a winner had been bought there. It isn't claiming that their tickets are more likely to win."

"Not in so many words, but it misleads people into thinking that's the case. That's what I don't like about it."

"Okay, I'll give you that."

"Lots of problems in probability go counter to intuition. You have to stick to the math to avoid being misled."

"And you have to stick to that soup until it's gone!" my mother said.

1.3 THE MONTY HALL PROBLEM

1.3.1 Three Doors

I somehow managed to finish my soup, then headed back to my room with Yuri.

"Speaking of probability," I said, "do you know the Monty Hall problem?"

The Monty Hall problem

Monty Hall stands before three doors and says,

> There's a prize behind just one of these doors, and nothing behind the other two. Which do you choose?

You pick one of the doors, but then Monty says,

> By the way, I know where the prize is. As a hint, I'll open one of the doors you didn't choose. Of course, I won't choose the one with the prize behind it. If you've already chosen the door with the prize behind it, I'll randomly choose one of the other two.

Monty opens one of the doors that you did not choose and, sure enough, there is nothing behind it.

> I'm also going to allow you to switch your choice, if you like. Will you stick with the door you first chose, or will you switch to the other unopened door?

If you want to win the prize, should you stick with the first door you chose, or should you switch?

"I really don't see myself ever being in this situation," Yuri said.

"Lots of people were, though," I said, "on a game show called Let's Make a Deal. Contestants would win, like, either a new car or a goat."

A Bet You Cannot Lose

"Oooh! I want the goat! Definitely the goat!"

"Uh, that's what the losers got."

"The winners, I'd say. And who's this Monty Hall guy?"

"The emcee of the game show."

"Can we trust him? He's not going to shuffle prizes around behind the doors after I've made my choice?"

"I don't think we have to worry about that."

"Well then what's the point in switching? It's a trick question, right? You're going to make me do lots of work, then say it doesn't really matter what I do."

"So that's your answer? That it doesn't matter if you switch?"

"Right! Whether I switch or not, I have the same probability of winning."

"Nope."

"No?"

"No, you should always switch doors."

"Prove it."

"Sure, just look at all the possibilities. Here, let's make a table, labeling the doors A, B, and C. Circles are winners, Xs are losers."

Probability		A	B	C	
1	$\frac{1}{3}$	O	×	×	A is the winner
	$\frac{1}{3}$	×	O	×	B is the winner
	$\frac{1}{3}$	×	×	O	C is the winner

"Makes sense," Yuri said, nodding.

"There's an equal $\frac{1}{3}$ probability of any of these three cases being true," I said, "and three ways of choosing for each of the cases. Here, let me make a new table. The brackets show which door you chose."

Probability			A	B	C	
1	$\frac{1}{3}$	$\frac{1}{9}$	[○]	×	×	A wins, A chosen
		$\frac{1}{9}$	○	[×]	×	A wins, B chosen
		$\frac{1}{9}$	○	×	[×]	A wins, C chosen
	$\frac{1}{3}$	$\frac{1}{9}$	[×]	○	×	B wins, A chosen
		$\frac{1}{9}$	×	[○]	×	B wins, B chosen
		$\frac{1}{9}$	×	○	[×]	B wins, C chosen
	$\frac{1}{3}$	$\frac{1}{9}$	[×]	×	○	C wins, A chosen
		$\frac{1}{9}$	×	[×]	○	C wins, B chosen
		$\frac{1}{9}$	×	×	[○]	C wins, C chosen

"Got it."

"There's $3 \times 3 = 9$ patterns in all, each with an equal probability of $\frac{1}{9}$. If you chose the prize on your first guess, the emcee has two ways of choosing which door to open. In that case, we split into two cases with total probability $\frac{1}{9}$, or $\frac{1}{18}$ each."

"Okay."

"But if you've chosen the wrong door, the emcee only has one way of choosing which other door to open, so the probability remains unchanged at $\frac{1}{9}$. I know, it's getting hard to follow again. One more table, breaking things down into probabilities of 1, $\frac{1}{3}$, $\frac{1}{9}$, and $\frac{1}{18}$."

A Bet You Cannot Lose

Probability				A	B	C	
1	$\frac{1}{3}$	$\frac{1}{9}$	$\frac{1}{18}$	[O]	⇒	×	A wins, A chosen, B opened
			$\frac{1}{18}$	[O]	×	⇒	A wins, A chosen, C opened
		$\frac{1}{9}$		O	[×]	⇒	A wins, B chosen, C opened
		$\frac{1}{9}$		O	⇒	[×]	A wins, C chosen, B opened
	$\frac{1}{3}$	$\frac{1}{9}$		[×]	O	⇒	B wins, A chosen, C opened
		$\frac{1}{9}$	$\frac{1}{18}$	⇒	[O]	×	B wins, B chosen, A opened
			$\frac{1}{18}$	×	[O]	⇒	B wins, B chosen, C opened
		$\frac{1}{9}$		⇒	O	[×]	B wins, C chosen, A opened
	$\frac{1}{3}$	$\frac{1}{9}$		[×]	⇒	O	C wins, A chosen, B opened
		$\frac{1}{9}$		⇒	[×]	O	C wins, B chosen, A opened
		$\frac{1}{9}$	$\frac{1}{18}$	⇒	×	[O]	C wins, C chosen, A opened
			$\frac{1}{18}$	×	⇒	[O]	C wins, C chosen, B opened

"This is starting to become kind of a pain," Yuri said, but peered closely at the table nonetheless. "And?"

"Well, the cases where your first pick was the winner are the ones with the bracketed circles."

Probability that first pick wins = Sum of probabilities of [O]

$$= \frac{1}{18} + \frac{1}{18} + \frac{1}{18} + \frac{1}{18} + \frac{1}{18} + \frac{1}{18}$$
$$= \frac{6}{18}$$
$$= \frac{1}{3}$$

"Told ya," Yuri said. "The probability of a win is $\frac{1}{3}$."

"Not so fast. To find the probability of winning after a switch, we sum up the probabilities of the bracketed Xs."

$$\text{Probability that switched door wins} = \text{Sum of probabilities of } [\,\times\,]$$
$$= \frac{1}{9} + \frac{1}{9} + \frac{1}{9} + \frac{1}{9} + \frac{1}{9} + \frac{1}{9}$$
$$= \frac{6}{9}$$
$$= \frac{2}{3}$$

"Oh."

"To sum up..."

$$\text{Probability that first pick wins} = \frac{1}{3}$$
$$\text{Probability that switched door wins} = \frac{2}{3}$$

"... so you're definitely better off switching."

"Mmm, I guess that makes sense," Yuri said, clasping her hands behind her head and frowning. "But still, I wish it clicked better."

"Maybe there's a better way to see it... How about this? Say that instead of three doors, there's ten thousand, but still only one prize."

"Not sure I'd bother playing in that case, but go on."

"So you choose one door, and Monty opens 9998 of the remaining 9999 doors. Would you switch doors in that case?"

"Well sure. I mean, I only have a $\frac{1}{10,000}$ chance of getting my first pick right. I'm practically guaranteed to lose!"

"That's right."

"That means there's a $\frac{9,999}{10,000}$ chance that the prize is behind a door I didn't choose. Monty knows where it is, though, and is sure not to open that door as he runs around opening 9,998 other ones. I don't know what the probability is, but the prize is almost definitely behind the one door he didn't open."

"Right. You can think of it as the $\frac{9,999}{10,000}$ probability of the prize being behind one of the doors you didn't choose being smooshed down to just the one remaining door."

"Like boiling down a whole pot of beans," Yuri said with a grin.

"Er, yeah. Anyway, it's the same situation when there are just three doors. You have a $\frac{1}{3}$ chance of choosing the prize at the beginning, or in other words a $\frac{2}{3}$ chance of choosing wrong. When the emcee shows you a wrong answer, he's smooshing that $\frac{2}{3}$ chance of a win into just one door."

"Huh, cool."

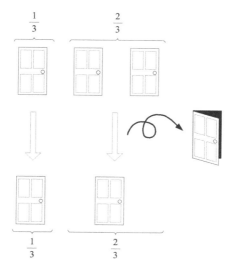

"So in the end," I said, "the Monty Hall problem just asks if you would rather hold on to the $\frac{1}{3}$ probability you started out with, or take the boiled down $\frac{2}{3}$ probability you're offered."

"Well why didn't you say so in the first place!"

1.3.2 A God's-Eye View

"I'm going to have to do more of this 'thinking with tables' thing," Yuri said.

"Good idea. Tables can give you a better view of the big picture. A god-like viewpoint, if you will."

"A god-like viewpoint? If I become a god, I'll be able to see everything? Even the future?"

"If you become an omniscient god, sure, I guess."

"So I would know the answer to the problem, and who I'll marry, and when I'll die? All of it?"

"Maybe you should just worry about solving problems in the mortal way for now, studying hard so you can do it yourself."

"You're no fun."

"You're in your last year of junior high now. Don't forget, you have high school entrance exams coming up."

"Now you're even less fun. Don't remind me about exams. Besides, you have your own exams coming up."

"Touché," I said, wincing. I didn't want to be reminded about them, either.

"I'm not all that worried about exams, honestly," Yuri said. "I'm more miffed that you and Miruka won't be there when I finally get to high school."

"Tetra will be there."

"Yeah, but only for one year."

"At least you'll have someone to talk to. That's kind of a big deal."

"I guess."

Yuri slipped into uncharacteristic silence, and I joined her in thought.

Someone to talk to... The friends I had made over the past couple of years had profoundly changed my life. When I first started high school, I never would have predicted how things had turned out. Meeting Miruka that year, Tetra the next. Friends who I could talk to, about math, sure, but also what I was feeling. We can never predict what will happen next, what life has in store...

"By the way," Yuri said, interrupting my musings.

"Huh?"

"There's something I've been meaning to ask you," she said, twirling a lock of hair and drawing out her words.

"Uh oh, she's getting serious."

"Have you, mmm..."
"Go on, spit it out."
"... ever kissed a girl?"

> "If all sides of a cube were identical, how could we tell which side is face up?"
>
> GRAHAM, KNUTH, & PATASHNIK
> *Concrete Mathematics* [8]

CHAPTER 2

Step By Step

> By this means all her quarter was free, and all that was in that part was dry. You may be sure my first work was to search, and to see what was spoiled and what was free.
>
> DANIEL DEFOE
> *Robinson Crusoe*

2.1 AT SCHOOL

2.1.1 Tetra

"Hey!" a chipper voice shouted, causing me to turn.

I had just finished my last class, and I was standing in the hallway, right under a sign that read 'Quiet in the Hall.'"

"Inside voice, Tetra," I said.

Tetra was a second-year student, one year below me. She was small, and cute, and sometimes had more energy than she could fully contain.

"Oops, sorry!" she whispered, a hand over her mouth.

Same old conversation, same old Tetra, same old— Wait.

Behind her was a girl I had never seen before, a girl with hair the color of flames. It reached down to her shoulders, and looked like it had been styled with hedge clippers. My first impression was 'wild animal.'

"Oh, this is Lisa!" Tetra said. "She just started here this year as a first year student."

I nodded awkwardly. It somehow felt strange for Tetra to not be the youngest in a group.

Lisa just stood there impassively, not acknowledging Tetra's introduction. Pretty, but expressionless.

"Nice to meet you," I said. "I'm sorry, that name was...?"

"Lisa," she said. "Lisa Narabikura."

2.1.2 Lisa

Moving to the library, Lisa, Tetra, and I sat in a line at a table. As soon as we'd settled in, Lisa had pulled out an ultrathin laptop—her computer was the same brilliant red as her hair, I noted—and hadn't stopped typing since. Nor had she uttered a word. At one point Tetra said, "So anyway...," causing Lisa to briefly glance up, but her eyes soon moved back to the screen. Her fingers kept pounding the keyboard the entire time.

"Narabikura, as in the Narabikura Library?" I said.

Lisa nodded silently.

"Oh, right!" Tetra said. "Lisa is Dr. Narabikura's daughter."

"Huh."

Despite being the topic of conversation, Lisa remained engaged with her computer.

"So how's it feel to be a second-year student?" I asked Tetra, wanting to break the uncomfortable silence.

"Good, I guess," she said. "A new school year puts me in the mood for starting new things."

"Like what?"

"I want to learn more about algorithms."

"Algorithms?" I said. "Like, computer algorithms?"

"Right!" Tetra said, brightening. "An algorithm is, uh, a specific procedure for obtaining a desired output from provided input. Usually programmed on a computer." She was clearly enjoying this. "Algorithms have several features. Input, output, a clear procedure, and ... um ... let's see. There were two more, but I forgot. Anyway, most important is that there's nothing unclear about the process."

"Cool," I said. I didn't know much about computers. I had read a book or two about programming, but it had never really clicked. "So you're studying all that?"

"Yeah. Well, I've been studying about computers in general for a while, so the other day Mr. Muraki suggested that I look into algorithms. Oh, and he gave me a card about linear searches."

"What's a linear search?" I asked.

Tetra grinned. "So glad you asked!"

2.1.3 Linear searches

"A linear search is a search algorithm, one where you look for a specific number in a sequence. Well, it doesn't have to be a number, I guess, but I'll just talk about that for now."

Tetra opened her notebook to a blank page.

"Say we have a sequence like this."

$$\langle 31, 41, 59, 26, 53 \rangle$$

"And let's also say that among these five numbers, you want to know if you can find a 26. Now, we humans can just glance at that and the 26 just jumps out, but a computer has to go through the numbers one by one. A linear search is an algorithm that says 'start at the beginning, and examine each number in order.' In this case, it would work like this."

$\langle \, \textcircled{31}, \ 41, \ 59, \ 26, \ 53 \, \rangle$ first number is 31, which isn't 26
$\langle \, 31, \ \textcircled{41}, \ 59, \ 26, \ 53 \, \rangle$ second number is 41, which isn't 26
$\langle \, 31, \ 41, \ \textcircled{59}, \ 26, \ 53 \, \rangle$ third number is 59, which isn't 26
$\langle \, 31, \ 41, \ 59, \ \textcircled{26}, \ 53 \, \rangle$ fourth number is 26, which is 26

                          Found it!

"That's ... kind of obvious."

"That's what I thought too, at first," Tetra said. "But remember—'It's always best to start from the obvious.'"

"Ah, now you're quoting me. Thank you for the reminder."

I had indeed said the same thing to Tetra several times in the past, while teaching her math. I had been tutoring her and helping her with math problems for two years, but now the tables were turned. It felt kind of good.

"I mean, what if you had, like, a million numbers!" Tetra threw her arms up. "You wouldn't want to comb through all that, would you? But a computer can do it. So long as it has a program—an algorithm—to do the search."

She pulled a card out of her bag.

"Here's a summary of the linear search algorithm."

The linear search algorithm (input and output)

Input
- Sequence $A = \langle A[1], A[2], A[3], \ldots, A[n] \rangle$
- Sequence length n
- Target value v

Output

If there exists a value equal to v in A
  Output $\langle \text{FOUND} \rangle$.
If there exists no value equal to v in A
  Output $\langle \text{NOT FOUND} \rangle$.

"We give the linear search a sequence $A = \langle A[1], A[2], A[3], \ldots, A[n] \rangle$ that has n values in it and a value v that we're looking for, and the algorithm tells us if that value is in the sequence. So the inputs are A, n, and v."

"Sure," I said, nodding. "And the $A[1]$ here is like A_1?"

"Right. I'm just writing it a different way, $A[1], A[2], A[3], \ldots, A[n]$ instead of $A_1, A_2, A_3, \ldots, A_n$. Anyway, $A[1]$ is element 1 in the sequence A."

"I get that."

"Given those inputs, if there's a number equal to v in sequence A, then we display $\langle \text{FOUND} \rangle$, and if no such number exists in A we display $\langle \text{NOT FOUND} \rangle$. So that's our output, $\langle \text{FOUND} \rangle$ or $\langle \text{NOT FOUND} \rangle$."

"I see. So the output is like the result."

"Yeah, sure! And this is pseudocode for the procedure the linear search algorithm uses."

Tetra pulled out another card.

The linear search algorithm (pseudocode)

$L1$:	procedure LINEAR-SEARCH(A, n, v)
$L2$:	$\quad k \leftarrow 1$
$L3$:	$\quad$ while $k \leqslant n$ do
$L4$:	$\quad\quad$ if $A[k] = v$ then
$L5$:	$\quad\quad\quad$ return $\langle \text{FOUND} \rangle$
$L6$:	$\quad\quad$ end-if
$L7$:	$\quad\quad k \leftarrow k + 1$
$L8$:	$\quad$ end-while
$L9$:	$\quad$ return $\langle \text{NOT FOUND} \rangle$
$L10$:	end-procedure

I cocked my head. "Pseudocode?"

"Yeah, a sorta-kinda computer program. Not in a real programming language that a computer could actually run, though. It just shows how an algorithm works."

"Right, got it."

"Mr. Muraki says that almost every book is different in how it describes algorithms. But that's okay, so long as it clearly shows every step it uses to turn inputs into output."

"Gotcha. And this is the linear search algorithm."

"Sure. In short, linear search is an algorithm that sequentially checks through $A[1], A[2], A[3], \ldots A[n]$ to see if there is a value equal to v in there."

"I should start reading this from line $L1$?"

Tetra nodded vigorously several times. "Exactly. Mr. Muraki said we have to execute each step, from $L1$ to $L10$."

"Execute?"

"Yeah..." Tetra spoke slowly as she started rereading the card. "We're supposed to pretend like we're a computer executing this program. We say, 'Hi, I'm a computer. Oh look, some input. Okay, I'll follow this algorithm line by line, and do exactly what it says.' It can be a little bit tedious, but Mr. Muraki says that's the best way to understand how an algorithm works."

"Huh."

"Honestly, it's the kind of grunt work I enjoy. Here, I'll show you!"

"Tetra 1.0, initiate," Lisa mumbled.

2.1.4 A Walkthrough

"Okay, this is called a walkthrough," Tetra said. "We're going to use some specific example input to walk through the linear search algorithm line-by-line. Here's the test case we'll use."

$$\begin{cases} A = \langle\, 31, 41, 59, 26, 53\,\rangle \\ n = 5 \\ v = 26 \end{cases}$$

"In other words, we'll use these values..."

$$A[1] = 31,\ A[2] = 41,\ A[3] = 59,\ A[4] = 26,\ A[5] = 53$$

"...and perform a linear search to look for the number 26."

Tetra moved the card with the algorithm to the middle of the table.

"Like you said, we start from line $L1$."

$\boxed{1}$ $L1$: procedure LINEAR-SEARCH(A, n, v)

"That line says that a procedure called 'LINEAR-SEARCH' starts here. It also tells us that this procedure takes A, n, and v as inputs. Okay, on to line $L2$."

$\boxed{2}$ $L2$: k ← 1

"This line stores 1 in a variable named k. So after this line is executed, k has the value 1."

"Why do we do that?" I asked.

"Because we're going to use k to keep track of which term in the sequence we're looking at."

"Gotcha. Okay, on to line $L3$."

$\boxed{3}$ $L3$: while k ⩽ n do

"Right!" Tetra said. "This line is called a repeat condition, in this case a 'while' condition. That's a keyword that says, 'keep repeating everything from here to the end-while line, so long as this condition is true.'"

"The condition being this k ⩽ n here?"

"Yep! The variable n is the length of the sequence, so k ⩽ n makes sure we don't go beyond the range of numbers we're interested in. Technically it would be 1 ⩽ k ⩽ n, I guess, but k starts with a value of 1 and gets bigger from there, so all we need to worry about is whether it's gotten bigger than n."

"Makes sense," I said, nodding.

"Great. Right now we have k = 1 and n = 5, so the condition k ⩽ n holds. That means we can go on to line $L4$."

$\boxed{4}$ $L4$: if A[k] = v then

"This is another conditional statement, an 'if.' It says that we can execute the following lines—up to the end-if statement—so long as its condition is met."

"And this time the condition is A[k] = v, which would mean that the number we're looking at equals the number we want."

Tetra nodded. "Right now k = 1, so we check the first number in the sequence. A[k] = A[1] = 31, so we compare A[k] = 31 and v = 26. They're different numbers, so the first number in the sequence isn't the number we're looking for. Since the condition doesn't hold, we skip everything until the end-if line. That takes us all the way down to line $L7$."

$\boxed{5}$ $L7$: k ← k + 1

"That line says to replace what's in variable k with the value k+1. Right now k = 1, so the value of k + 1 is 2. That means k increases from 1 to 2. Next, we move on to line $L8$."

$\boxed{6}$ $L8$: end-while

"The end-while in this line is paired with the 'while' statement in line $L3$, so we jump back up there."

"Wow, Tetra," I said. "You're really getting this stuff down."

Tetra blushed and looked down. "Nah, not really."

"Seriously, you do a great job of explaining all this, jumping back and forth between form and meaning. You're able to discuss both how the program is written and what the algorithm is doing."

"Uh, I guess."

"Still, all this seems like so much work," I said, looking through Tetra's notebook. It was filled with details about writing pseudocode.

"It can be, but once you get used to keeping track of all the variable values as you go through each line, it's not so bad."

"Continue," Lisa said.

"Oh, right," Tetra said. "Next line."

| 7 | $L3$: while $k \leqslant n$ do

"We've gone back to line $L3$, so we have to check our repeat condition again. Now $k = 2$ and $n = 5$, so the condition $k \leqslant n$ still holds. On to line $L4$."

| 8 | $L4$: if $A[k] = v$ then

"Another condition check. Now the value of variable k equals 2, so the condition $A[k] = v$ doesn't hold, because the value of the input $A[2]$ is 41 and v is 26. That means the second value in the sequence isn't the number we're after, either."

"And that means we skip line $L5$ and line $L6$, and jump to line $L7$, right?"

| 9 | $L7$: $k \leftarrow k + 1$

"Right! That's where we increase variable k by 1, so now it becomes 3. On to line $L8$..."

| 10 | $L8$: end-while

"...which takes us back to line $L3$."

"Lather, rinse, repeat," I said.

"Yes," Tetra said, "but with k increasing each time."

"Continue," Lisa said.

| 11 | $L3$: while $k \leqslant n$ do

"Okay, now $k = 3$," Tetra said, "so again $k \leqslant n$ holds. We can move on to line $L4$."

| 12 | $L4$: if $A[k] = v$ then

"Now k = 3, so A[k] = v doesn't hold, since the value of input A[3] is 59. Jump to line *L7*."

|13| *L7*: k ← k + 1

"Now the value of variable k is 4. On to line *L8* ..."

|14| *L8*: end-while

"... which sends us once again to line *L3*."

"I'm getting déjà vu here," I said.

"Except that k has increased again."

"Continue," Lisa said.

|15| *L3*: while k ⩽ n do

"Now k = 4, so k ⩽ n still holds and we advance to line *L4*."

|16| *L4*: if A[k] = v then

"Now k = 4, and look! The condition A[k] = v holds, because the value of A[4] is 26! At last, we've met the condition of the 'if' statement. On to line *L5*."

|17| *L5*: return ⟨FOUND⟩

"This 'return' represents the output of the procedure, which in this case is the result ⟨FOUND⟩. We jump to the end-procedure in line *L10*."

|18| *L10*: end-procedure

"And that's it! The LINEAR-SEARCH procedure is complete."

"Took long enough," I joked.

"Yeah, 18 steps with the input $A = \langle 31, 41, 59, 26, 53 \rangle, n = 5, v = 26$, but we're done now! And we got a ⟨FOUND⟩ output!"

"Somehow it seems like a lot of work, just to find the one number."

"Yeah, overblown somehow, what with all that fiddling with the k variable."

"Makes it hard to see exactly the path we took."

"Well, I guess we could show things more compactly like this."

```
L1:   procedure LINEAR-SEARCH(A, n, v)
L2:       k ← 1
L3:       while k ⩽ n do
L4:           if A[k] = v then
L5:               return ⟨FOUND⟩
L6:           end-if
L7:           k ← k + 1
L8:       end-while
L9:       return ⟨NOT FOUND⟩
L10:  end-procedure
```

Walkthrough for the linear search algorithm
(with inputs $A = \langle 31, 41, 59, 26, 53 \rangle, n = 5, v = 26$).

"Much clearer," I said.

"That's what's so cool about computers! They can do things like this all day without getting bored!"

"I'd say you just did a pretty good job of that yourself."

"CompuTetra," Lisa said.

2.1.5 Analyzing the Linear Search

"So Mr. Muraki gave you this card as a research topic?" I said.

"Looks that way," Tetra said, nodding.

Mr. Muraki had been giving us these cards since I'd started high school. Sometimes they seemed like simple problems—solve for x, that kind of thing—but often it was just some mathematical *something* where it wasn't even obvious what the problem was. Common among these cards was that they were very different from the kind of thing we did in math class. They were triggers leading us to create our own problems, and to solve them on our own, using our own heads and hands. Mr. Muraki's assignments weren't just about finding solutions to problems; they showed us how to discover our own problems.

Even so, Tetra's card felt different from the kind of math we had tackled before. I wasn't sure where the math was, beyond the ridiculously simple equations like $k \leqslant n$ and $A[k] = v$.

"I think I get linear searches," I said, "but where do we go next? I'm not sure what kind of problem to create from a card like this."

I cast a sideways glance at Lisa, who was still tapping away. She hadn't taken her hands off of her keyboard the entire time.

"Good question," Tetra said, blinking her eyes. "Maybe we can find a way to make this algorithm go faster? The whole point of an algorithm is to produce some output, so it seems like faster would be better."

I nodded.

"That makes sense. But the whole point of the linear search algorithm is to comb through things one at a time, so is that even possible? Besides, what would 'faster' even mean for a bunch of steps written down on paper?"

"Uh, also good questions," Tetra said.

"Number of executions," Lisa said. She gave me an expressionless look, her fingers still dancing across her keyboard.

"Meaning what, exactly?" I said.

"Count," the mysterious redhead replied. Something about the huskiness of her sparse words against the *tap-tap-tap* of her constant typing was oddly attractive, and added weight to what she said.

"Count the number of lines we executed, you mean?" Tetra asked. "Let's see, end-procedure came at $\boxed{18}$, so I guess the number of executions was 18."

"In this case it was," I said.

"What do you mean?"

"Well, depending on the input you might find v in sequence A, or you might not. Even if you do, it can be at the start of the sequence, or at the end, or anywhere in-between. It's kind of hard to break things down into so many possible cases."

"Oh."

"And we haven't even brought up the input n yet. Like you said, n could be one million. So I'm not sure how we count steps for all cases of a v that may not even be there."

"Me neither," Tetra sighed.

2.1.6 Analysis of Linear Search When v is in the List

Lisa silently spun her computer around so that it faced us. On its screen was a table listing the steps in the linear search algorithm, paired with symbols like 1 and M in a column labeled "Executions."

	Executions	Linear Search
$L1$:	1	procedure LINEAR-SEARCH(A, n, v)
$L2$:	1	$k \leftarrow 1$
$L3$:	M	while $k \leqslant n$ do
$L4$:	M	if $A[k] = v$ then
$L5$:	1	return $\langle$ FOUND $\rangle$
$L6$:	0	end-if
$L7$:	$M - 1$	$k \leftarrow k + 1$
$L8$:	$M - 1$	end-while
$L9$:	0	return $\langle$ NOT FOUND $\rangle$
$L10$:	1	end-procedure

Number of executions when v is found.

"What's this M?" Tetra asked her.

"The position of v."

"Oh, I get it," I said, examining the table. "It's the number of times each line gets executed."

This was starting to look like math after all—Lisa was using variables in place of an unknown, which in this case was the number of times a given line in the algorithm would be executed. This M was something I did all the time, generalizing through the introduction of a variable.

"What do we do with this?" Tetra asked.

"Add them up!" I said. "By adding the number of executions for each line, we can find the total number of lines executed! We just need to create a formula that contains this variable M."

Number of executed steps (v found)
$$= L1 + L2 + L3 + L4 + L5 + L6 + L7 + L8 + L9 + L10$$
$$= 1 + 1 + M + M + 1 + 0 + (M - 1) + (M - 1) + 0 + 1$$
$$= M + M + M + M + 1 + 1 + 1 + 1 - 1 - 1$$
$$= 4M + 2$$

"So we can produce the output in $4M + 2$ steps?" Tetra said. "Assuming that v is in there, at least?"

"That's right. For example, in your test case we were looking for 26, so—"

"Wait, wait!" Tetra shouted. "We can just calculate it out, right? I've got this!"

"Be my guest," I said. Tetra and I were both old hands at what came next. Lisa spun her computer back around, and began typing again.

"Okay, in the test case we were looking for 26 in $\langle 31, 41, 59, 26, 53 \rangle$. The 26 is in the fourth position, so $M = 4$."

Number of executions in the test case

$= 4M + 2$
$= 4 \times 4 + 2$ substitute $M = 4$
$= 18$ calculated

"Nice," I said.

"Sure enough, we ended on step $4M + 2 = 18$."

"Confirming that the number of executed steps is $4M + 2$, *if v exists*. Which naturally raises the question—"

"—what if it doesn't?"

> **Problem 2-1 (Execution steps in a linear search)**
>
> Find the number of executed steps in the linear search algorithm in the case where value v is not found in sequence $A = \langle A[1], A[2], A[3], \ldots, A[n] \rangle$.

2.1.7 Analysis of Linear Search When v is Not in the List

Lisa turned her computer back around and showed us her edits.

	Executions	Linear Search
L1:	1	procedure LINEAR-SEARCH(A, n, v)
L2:	1	$k \leftarrow 1$
L3:	$n+1$	while $k \leqslant n$ do
L4:	n	if $A[k] = v$ then
L5:	0	return $\langle$FOUND$\rangle$
L6:	0	end-if
L7:	n	$k \leftarrow k+1$
L8:	n	end-while
L9:	1	return $\langle$NOT FOUND$\rangle$
L10:	1	end-procedure

Number of executions when v is not found.

"Look! The M's are gone!" Tetra said.

"And now they're just n's? Does that work?"

I sat back to think through this. I could see why lines L1 and L2 would be executed once each. But line L3? Why would that be $n+1$, not just n?

After a few moments my doubts cleared. It had to be $n+1$, because $k \leqslant n$ would hold n times, namely once for each of $k = 1, 2, 3, \ldots, n$. Then there would be the one time where it did not hold, when $k = n+1$. Together, that line would execute $n+1$ times, giving the value for line L3. I nodded, impressed at how Lisa was able to just whip that out.

"$L3 = L2 + L8$," Lisa said.

I paused, not sure exactly what she meant, but carried on.

So how about line L4? If we don't find the value v, then we should have to compare each of the n values $A[1]$ through $A[n]$ with v, so it made sense that line L4 would be executed n times.

Line L5... Well, we're never going to output $\langle$FOUND$\rangle$, so lines L5 and L6 should be 0, sure.

Lines L7 and L8 should both be executed n times, for the same reason as line L4. And of course lines L9 and L10 would execute once each, when the the algorithm completed with the output $\langle$NOT FOUND$\rangle$. So everything checked out.

I glanced at Tetra, and saw that she was writing an equation.

Number of executed steps (v not found)
$= L1 + L2 + L3 + L4 + L5 + L6 + L7 + L8 + L9 + L10$
$= 1 + 1 + (n+1) + n + 0 + 0 + n + n + 1 + 1$
$= n + n + n + n + 1 + 1 + 1 + 1 + 1$
$= 4n + 5$

"That should be all the cases," Tetra said.

$$\text{Executed steps in a linear search} = \begin{cases} 4M + 2 & \text{when found} \\ 4n + 5 & \text{when not found} \end{cases}$$

Indeed. There was something comforting about seeing this as a mathematical representation. It let me think of this computer problem as something more like a math problem, something I hadn't realized was even possible. An hour ago, I would have said that there was no overlap between math and computer programs. Now, I was no longer so sure.

Solution 2-1 (Execution steps in a linear search)

If value v is not found in sequence $A = \langle A[1], A[2], A[3], \ldots, A[n] \rangle$, the linear search algorithm executes $4n + 5$ steps.

2.2 Analysis of Algorithms

2.2.1 Miruka

"Yikes!"

Lisa, who had been so cool and collected up to that point, squealed in a suddenly clear voice.

"Hi, Lisa. It's been a while," came a silky response.

I turned to see long black hair. Metal-framed glasses. Fingers that moved like a conductor's.

It was Miruka, formidably bright classmate and resident math genius. I had been studying math with her since we first met two years before, amidst a flurry of cherry blossoms. Actually, better to say that I'd been trying to keep up with her for two years. She swam in a mathematical sea far deeper and broader than the one I knew. She was our guide in our mathematical journey.

But that wasn't all. I... Well, let's just say that it had become painful to look at her, knowing that this would be our last year in school together. When we graduated high school, she would—

No, enough of that.

"Cut it out!" Lisa was saying. Miruka was tousling her hair. Lisa coughed as she tried to bat Miruka's hand away.

"What'cha up to, Miss Kurabikura?"

2.2.2 Characteristics of Good Algorithms

"Analysis of algorithms, huh?" Miruka said, peering at the computer screen over Lisa's shoulder. Lisa nodded. "Counting executed steps is certainly a fine start, but—"

Miruka's pause caused Lisa to look up.

"—there are some prerequisites you need to clarify if you want to find true execution times," Miruka continued. "Like, how long it takes to execute a given step. You can't say what's fast and what's slow without knowing that."

Good point, I thought.

"To do that you need a calculation model," Miruka said. "Lisa's model assumes that every step requires the same time to run. Like, that '$k \leftarrow 1$' and 'if $A[k] = v$ then' finish executing in the same amount of time. It has the advantage of simplicity, I'll give it that."

Tetra chimed in. "Miruka! Do you know the necessary features of an algorithm? I remember that they need input, output, and a clear procedure, but—"

"Input, output, clarity, feasibility, finiteness. Though, strictly speaking, there may be no input."

"By clarity, you mean it should be clear how the procedure operates, right? What about feasibility?"

"That it should be possible to actually implement the algorithm."

"Oh, okay. And finiteness?"

"That it will complete in a finite number of steps."

Tetra finished scratching in her notebook.

"Input, output, clarity, feasibility, finiteness. Got it!"

2.2.3 Eliminating Cases

Miruka skimmed over Tetra's notes.

$$\text{Executed steps in a linear search} = \begin{cases} 4M + 2 & \text{when found} \\ 4n + 5 & \text{when not found} \end{cases}$$

"Hmph."

"We were calculating the number of execution steps by case," Tetra said.

Miruka closed her eyes, and we all held our breath. Even the normally flighty Tetra fell silent, and Lisa retreated back behind her computer. After a time, Miruka wagged an index finger side to side, and opened her eyes.

"So...," she began, smiling. "You divided the linear search algorithm into two cases, one where v was in sequence A, and one where it wasn't. That's not wrong, but there's a better way. You can treat both cases simultaneously."

"Like, as a single case? But how?" I said.

"Yeah," Tetra said, clenching one hand into a fist. "I mean, we find v in one case, and we don't in the other. The two don't seem very combinable. They even have different outputs."

"Right," I said. "That's why we separated them in the first place."

Miruka sidled up to Lisa and whispered something into her ear. Lisa made an unpleasant face, but turned to her computer and started typing.

"Not a problem, we do it like this," Miruka said. Right on cue, Lisa spun her red laptop around to face us.

	Executions	Linear search
$L1$:	1	procedure LINEAR-SEARCH(A, n, v)
$L2$:	1	$k \leftarrow 1$
$L3$:	$M + 1 - S$	while $k \leqslant n$ do
$L4$:	M	if $A[k] = v$ then
$L5$:	S	return $\langle$ FOUND $\rangle$
$L6$:	0	end-if
$L7$:	$M - S$	$k \leftarrow k + 1$
$L8$:	$M - S$	end-while
$L9$:	$1 - S$	return $\langle$ NOT FOUND $\rangle$
$L10$:	1	end-procedure

Number of executions, combining cases of when v is found and when it is not.

"What's this new variable, S?" Tetra asked nervously.

"'Generalization through introduction of a variable,'" Miruka quoted. "The whole point of generalization is to combine multiple special cases into one, right? We're going to define this S so that it can handle both cases."

· $S = 1$ means that v was found.
In this case, M equals the position of v.

· $S = 0$ means that v was not found.
In this case, M equals n.

"Why did you name it S?" Tetra asked.

"It doesn't really matter what you name variables, but here I'm using 'S' for 'success.' A successful find, that is. And I'm assigning to S a one-bit mapping to show the truth or falseness of the proposition 'v was found in sequence A.'"

$$v \text{ was found in sequence } A \quad \Longleftrightarrow \quad S = 1$$
$$v \text{ was not found in sequence } A \quad \Longleftrightarrow \quad S = 0$$

"Interesting," I said. "So when S is 1 the result is ⟨FOUND⟩, and when it's 0 the result is ⟨NOT FOUND⟩."

"More to the point, by adding a variable we've gotten rid of the separate cases," Miruka said.

"Which means you're able to represent the number of execution steps in a linear search using a single equation!" I said.

Tetra picked up her pencil and started calculating.

Execution steps in a linear search

$= L1 + L2 + L3 + L4 + L5 + L6 + L7 + L8 + L9 + L10$

$= 1 + 1 + (M + 1 - S) + M + S + 0 + (M - S) + (M - S) + (1 - S) + 1$

$= 4M - 3S + 5$

2.2.4 Considering Meaning

Tetra looked up with a serious expression and said, "Yep, the $4M - 3S + 5$ calculation works out. But still... maybe it's just me, but this variable S seems kinda, I don't know, overly convenient. Is it really okay to use a variable like this?"

"It is," Miruka said.

"Sure," I said. "There's nothing vague about it, and it doesn't contradict anything. All we've done is define a variable with a specific value."

Miruka nodded. "You'll have more fun with this if you stop worrying about how many variables are there, and focus on what they mean."

"What they mean?" Tetra said, scrunching her face.

"You. Scoot," Miruka said, waving me out of my seat next to Tetra.

I sighed, but acquiesced to the inevitable and moved to another chair.

"Okay, Tetra, answer me this," Miruka said. "What is M when $S = 1$?"

"Uh, M is the position of v, the number we're looking for."

"That's strictly inaccurate."

"Huh?" Tetra said.

"Huh?" I parroted.

Lisa remained silent.

"Say you're searching for $v = 26$ in the sequence $\langle 31, \underline{26}, 59, \underline{26}, 53 \rangle$. What then?"

"Ah, right. Then the v you're looking for is in more than one place."

"Indeed. By saying that M is the position of v, you're assuming there can only be one v in the sequence. So to be fully accurate, you have to say that M is the *least* value for positions of v."

"Getting kind of wordy," I said.

"Maybe," Miruka said. "But correct beats concise."

"Which is a good thing," Tetra said.

"Next question," Miruka said to Tetra. "What is S?"

"Oh, we talked about that before. S is a variable representing whether we've found v in the sequence."

"Correct. To introduce a technical term, S is an 'indicator,' a variable or equation that represents the truth or falseness of a proposition."

"Indicator..." Tetra said, writing in her notebook.

"Next question. What is $1 - S$?"

"Hmm, let's see... Oh! It's 0 if we find v, and 1 if we don't! Because it equals 1 when $S = 0$, and 0 when $S = 1$. It, like, flips back and forth between 1 and 0."

Tetra demonstrated this by flipping her hand palm up, palm down.

"Well done," Miruka said. "$1 - S$ is an indicator for 'v was not found.'"

"Yay! Another indicator!"

"Next question. What is $M + 1 - S$?"

Tetra consulted her notes, and I sat back to think.

· If $S = 1$, then $M + 1 - S$ equals M, which is the position of v (more precisely, the smallest of the positions of v).

· If $S = 0$, then $M + 1 - S$ equals $M + 1$. Which means...

...what?

"If $S = 1$, then $M + 1 - S$ is the position of v," Miruka said. "So what about when $S = 0$?"

"The next position of v, maybe?" I said.

"I don't think that's right," Tetra said. "If $S = 0$, that means we didn't find a v."

Oops.

"Yeah, you're right. Hmm..."

"So what does $M + 1 - S$ represent when $S = 0$?" Miruka prodded.

"$n + 1$," Lisa muttered.

Miruka turned toward her.

"Exactly. When $S = 0$, M equals n. So $M + 1 - S$ and $n + 1$ are the same."

"Uh, sorry, just what are we doing?" Tetra said. "I'm getting a little bit lost here."

"Hmph."

Miruka stood and started walking slowly around the table. A soft spring breeze blew in through the window, fluttering the hair of the math girls surrounding me. A citrus scent wafted by.

"This $M + 1 - S$ is quite interesting, actually." Miruka said as she paced. "When $S = 1$, it gives us the position of v. When $S = 0$, it's equal to $n + 1$. So how can we combine the two? In particular, is there some way we can consider $M + 1 - S$ as equal to the position of v in *all* cases?"

"But, Miruka," I said, "if $S = 0$, then v—"

"I know, I know. If $S = 0$, v isn't in $A1, A[2], A[3], \ldots A[n]$. So let's just force v into, say, $A[n + 1]$."

"*Force* it there?" I said.

"Sure. Then we can say that $M + 1 - S$ is always equal to the position of v." Miruka was speaking matter-of-factly, but I was just confused. "We're going to consider $M + 1 - S$ as a single thing, not as some split identity."

"A single thing..." Tetra whispered.

Something Miruka had said long ago came back to me.

What we thought were two things, we now see as one.

"A sentinel," Lisa said.

"Exactly so, Lisa," Miruka said. "Come over here."

"No."

"Um, what's a sentinel?" Tetra asked.

2.2.5 Sentinel Linear Search

Miruka whispered something in Lisa's ear, and Lisa went back to her keyboard. I was amazed at how quickly, and how quietly, she could type.

After a brief session of silent typing, Lisa spun her computer around to show something called a sentinel linear search.

Sentinel linear search (procedure)

$S1$:   procedure SENTINEL-LINEAR-SEARCH(A, n, v)
$S2$: $A[n+1] \leftarrow v$
$S3$: $k \leftarrow 1$
$S4$:   while $A[k] \neq v$ do
$S5$: $k \leftarrow k+1$
$S6$:   end-while
$S7$:   if $k \leqslant n$ then
$S8$:   return $\langle \text{FOUND} \rangle$
$S9$:   end-if
$S10$:   return $\langle \text{NOT FOUND} \rangle$
$S11$: end-procedure

Tetra and I stared at the computer screen. After a time, Tetra said "I can't figure this out without writing!" and reached for her notebook.

"Good idea," Miruka said. "Work through a test case."

```
S1:   procedure SENTINEL-LINEAR-SEARCH(A, n, v)
S2:       A[n + 1] ← v
S3:       k ← 1
S4:       while A[k] ≠ v do
S5:           k ← k + 1
S6:       end-while
S7:       if k ⩽ n then
S8:           return ⟨FOUND⟩
S9:       end-if
S10:      return ⟨NOT FOUND⟩
S11:  end-procedure
```

1
2
3

4	7	10	13
5	8	11	
6	9	12	

14
15

16

Walkthrough of the sentinel linear search
(Input: $A = \langle 31, 41, 59, 26, 53 \rangle, n = 5, v = 26$).

"I noticed something different about this walkthrough," Tetra said. "We repeat lines $S4 \to S5 \to S6$ over and over again. The way we check conditions is a lot simpler in this kind of linear search. But where's the sentinel?"

"It's the number assigned to $A[n+1]$ in line $S2$," Miruka said. "Storing v there will force the search to end, since once we get there we'll always 'find' the value at $k = n + 1$. So it's a 'sentinel' in the sense that it prevents the search from unintentionally straying too far off. In terms of code, it alleviates the need for range-checking k in the while loop at line $S4$."

"The LINEAR-SEARCH procedure required 18 steps to complete, but SENTINEL-LINEAR-SEARCH only needs 16, so I guess this one is a little faster. But still, just two steps?"

"In this case. But you have to think more generally, in terms of the number of steps when M is the execution count for line $S4$ and S is an indicator in the ⟨FOUND⟩ case."

	Executions	Sentinel linear search
S1:	1	procedure SENTINEL-LINEAR-SEARCH(A, n, v)
S2:	1	$A[n+1] \leftarrow v$
S3:	1	$k \leftarrow 1$
S4:	$M + 1 - S$	while $A[k] \neq v$ do
S5:	$M - S$	$k \leftarrow k + 1$
S6:	$M - S$	end-while
S7:	1	if $k \leqslant n$ then
S8:	S	return ⟨FOUND⟩
S9:	0	end-if
S10:	$1 - S$	return ⟨NOT FOUND⟩
S11:	1	end-procedure

Execution steps in sentinel linear search.

· $S = 1$ indicates the case where v was found.
In this case, M is the position of v.

· $S = 0$ indicates the case where v was not found.
In this case, M equals n.

Execution steps in sentinel linear search
$= S1 + S2 + S3 + S4 + S5 + S6 + S7 + S8 + S9 + S10 + S11$
$= 1 + 1 + 1 + (M + 1 - S) + (M - S) + (M - S) + 1 + S + 0 + (1 - S) + 1$
$= 3M - 3S + 7$

Tetra looked through the new calculation
"The original linear search required $4M - 3S + 5$ execution steps," she said, "but sentinel linear search only needs $3M - 3S + 7$!"

Procedure	Executed steps
LINEAR-SEARCH	$4M - 3S + 5$
SENTINEL-LINEAR-SEARCH	$3M - 3S + 7$

"Interesting!" I said. "We can use these execution step counts to compare how fast the algorithms run."

"And if we're comparing numbers, we can write it as an inequality!" Tetra said.

$$\text{Steps for linear search} > \text{Steps for sentinel linear search}$$
$$4M - 3S + 5 > 3M - 3S + 7$$

"Nice," I said. "Even better, we can subtract the right side from the left, and use the result to see what we gain from this."

Linear search execution steps − Sentinel linear search execution steps
$$= (4M - 3S + 5) - (3M - 3S + 7)$$
$$= 4M - 3S + 5 - 3M + 3S - 7$$
$$= M - 2$$

"We get a speedup when $M - 2$ is greater than 0," I said, again relieved to be back in the realm of equations. "So we want $M - 2 > 0$, in other words $M > 2$. So when we're searching through sequence A for a value v, sentinel linear search will be faster if the first one pops up in the third element or later."

"I love it when equations teach us things like that!" Tetra said.

"The worst-case scenario is when we get a ⟨NOT FOUND⟩ result. In that case linear search will need to execute $4M - 3S + 5 = 4n + 5$ steps to complete, while sentinel linear search will need $3M - 3S + 7 = 3n + 7$ steps."

2.2.6 *Creating History*

Miruka snagged my pencil and started twirling it around a finger.

"So by using sentinel linear search, we only have to execute about $\frac{3}{4}$ the execution steps as compared to the vanilla linear search," she said.

"Where do you get the $\frac{3}{4}$ from?" Tetra asked.

"The ratio of the M's. If M is sufficiently large, we can consider $4M - 3S + 5$ as $4M$, and $3M - 3S + 7$ as $3M$."

"Around 25% faster," Lisa said.

"Clarifying preconditions and obtaining the number of execution steps of an algorithm allows us quantitatively evaluate it. Quantitative evaluations are important because they let us say things like 'this algorithm is around 25% faster' instead of just 'it's faster.' Quantitative evaluations provide a basis for evaluating how good an algorithm is."

"Interesting," I said.

"Quantitative evaluations with clear conditions have other benefits. They're straightforward, and not open to interpretation. They help us verify the algorithm, and improve it. They can also be applied to the analysis of other algorithms."

"Quantitative evaluations with clear conditions...," Tetra said, scribbling in her notebook. "This is kind of like creating history, isn't it. I mean, an evaluation like that will live on, beyond the life of whoever did it. It gets passed on to the future, as a contribution to humanity."

"Wow, Tetra," I said, moved by her sentiment.

"You have to be careful, though," Miruka said, raising a finger. "Microscopic observations that pay too much attention to trivial differences between algorithms can cause you to miss some important commonalities. That's why asymptotic analysis is important. When it's large—"

"—linear search is order n," Lisa said.

"Big words for a little girl," Miruka snapped. "Do you know what they mean?"

Lisa shrugged and sank back.

"Just repeating jargon you've picked up, huh?"

Lisa silently glared at Miruka.

Wow, what's going on here? I wondered.

"Um, so anyway..." said a shaken Tetra.

After a time, Lisa clicked her tongue and looked away. I empathized with her, knowing I could never persevere in a stare-off with Miruka.

"The library is closed!" came a booming voice from behind us— the school librarian, Ms. Mizutani, making her punctual announcement that it was time to leave. I turned to see her give us a glance, then disappear back into her office.

2.3 AT HOME

2.3.1 A Step Forward

That night, I spent some time reflecting on the day. I thought I was getting a grasp on the basics of algorithms, as a finite number of clearly defined steps that turned input into output. I thought back on Tetra's careful walkthrough of the linear search algorithm, how well she persisted at precisely following what it did, and how this led to truly understanding it.

I recalled how using a formula to represent the number of steps an algorithm executed allowed us to analyze it, and what Miruka had taught us about quantitative evaluations and using variables to combine cases. I was impressed by how a close analysis of what seemed like a very straightforward algorithm revealed interesting discoveries. Such is the power of mathematics. It was *math* that allowed quantitative evaluation. It was writing things out mathematically that allowed evaluation, and comparison, and judgment.

I also thought about that redhead, Lisa—speedy silent typist and daughter of Dr. Kurabikura. She had known about sentinels and all that stuff, so clearly she was another motivated self-learner. She too realized that school doesn't really teach you much, that you have to study on your own. That we are our own best teacher.

Miruka, Tetra, and now Lisa. Each of them amazing in their own way. Compared to them, I—

No, cut that out!

I took off my glasses, and put a hand on my left cheek as I recalled my promise to Miruka.

I'm in my last year of high school. Soon, I'll be in college. I'll be able to truly study, to really accomplish something.

My own studies would be facing a quantitative evaluation soon, in the form of college entrance exams. I would be assigned a very weighty one-bit indicator: 1 for pass, 0 for fail.

With that thought I put my glasses back on, opened my notebook, and prepared to take another step forward.

> Very few people ever have a chance to choose the name for their life's work. But in the 1960s it was necessary for me to invent the phrase "analysis of algorithms," because none of the existing terms were appropriate for the kinds of things I wanted to do.
>
> DONALD KNUTH
> *Selected Papers on Analysis of Algorithms*

Tetra's notes

Pseudocode

Defining procedures

```
procedure <procedure name><variable array>
   <statement>
      ⋮
   <statement>
end-procedure
```

Defines a procedure using a procedure name and an array of variables passed as input.

Substitution

$$\langle variable \rangle \leftarrow \langle statement \rangle$$

Replaces the value of <variable> with the result of <statement>

Substitution (exchanging values)

$$\langle variable\ 1 \rangle \leftrightarrow \langle variable\ 2 \rangle$$

Swaps the values of <variable 1> and <variable 2>

"if" statement (1)

```
if <condition> then
    <processing>
end-if
```

1. Check if <condition> holds
2. If so, perform <processing>, then go to end-if
3. If not, go to the line after end-if

"if" statement (2)

```
if <condition> then
    <processing 1>
else
    <processing 2>
end-if
```

1. Check if <condition> holds
2. If so, perform <processing 1>, then go to "end-if"
3. If not, perform <processing 2>, then go to "end-if"

Note that one of <processing 1> or <processing 2> will always be performed.

<u>"if" statement (3)</u>

```
if <condition A> then
    <processing 1>
else-if <condition B> then
    <processing 2>
else
    <processing 3>
end-if
```

1. Check if <condition A> holds
2. If so, perform <processing 1>, then go to "end-if"
3. If not, check if <condition B> holds
4. If so, perform <processing 2>, then go to "end-if"
5. If neither <condition A> nor <condition B> hold, perform <processing 3>, then go to "end-if"

Note that one of <processing 1>, <processing 2>, or <processing 3> will always be performed.

"while" statement

```
while <condition> do
   <processing>
end-while
```

1. Check if <condition> holds.
2. If so, perform <processing>, then go to the "end-while" line, then return to the "while <condition> do" line.
3. If <condition> does not hold, go to the "end-while" line.

"return" statement

```
return <expression>
```

1. Find the value of <expression>, and take that as the resulting output of the procedure.
2. Go to the "end-procedure" line, and halt procedure execution.

CHAPTER **3**

The Solitude of 17,179,869,184

> This time I found much employment, and very suitable also to the time, for I found great occasion for many things which I had no way to furnish myself with but by hard labour and constant application.
>
> DANIEL DEFOE
> *Robinson Crusoe*

3.1 PERMUTATIONS

3.1.1 At the Book Store

Hands covered my eyes from behind.

"Guess who!" shouted a voice.

It was a Saturday, and I was in a large bookstore that had just opened next to the train station nearest my house. I was impressed so far—it had reading chairs scattered about, perfect for previewing books before deciding what to buy.

"That's not hard to figure out, Yuri," I said, peeling her hands from my face.

"You're no fun at all."

I turned to see her chestnut hair in a ponytail beneath a baseball cap.

"Are you here to buy books?" I asked.

"Why else would I be at a bookstore? But let's hang out up on the roof first."

"On the roof? But, these books..."

In the end, of course, I gave in to the inevitable and followed Yuri.

3.1.2 Getting It

Yuri went to the fence that surrounded the roof and looked down on the street below.

"Look at all the people down there!" she said.

I handed her a drink I'd bought at a convenient vending machine.

"Thanks!" she said. She popped it open and took a deep gulp.

I had hemmed and hawed past an answer to her awkward question the other day—whether I had ever kissed a girl—but still...

"Remember how you used to not be able to pronounce 'animal?'" I said.

"Oh, come on. I was, like, a little baby back then."

"Say it ten times fast, then."

"Easy. Animal, animal, animal, aminal— Oops."

"What was that?"

"Animal, animal, aminal— *Grrr*. Just...leave me alone!"

I laughed and apologized. It was good to see her back to her old self.

"By the way," she said, "do you know anything about permutations?"

"A little, I guess. Where did that come from?"

"It's just, there's something I don't get about them."

"How hard can junior high permutations be? It's stuff like lining up cards, right?"

> **Problem 3-1 (permutations)**
>
> In how many ways can you arrange four cards?
>
> $\boxed{A}, \boxed{B}, \boxed{C}, \boxed{D}$

"Finding the answer is easy!" Yuri said. "What I don't get is how my teacher explained it. He just goes straight to calculations. Arranging numbers, and people, and goats—"

"I doubt you're arranging goats."

"But it's all the same, yeah? It doesn't matter what you're arranging. I want to know the *why*, not the *how*."

"Okay, I think I see what you're after. Not what you plug into the equations, but where the equations themselves come from. You want an explanation that helps you *get it*, right?"

Yuri nodded and took a long swig from her drink. "That pretty much sums it up. Doing calculations over and over again is even more of a drag when you don't know what they mean."

"Something you have made quite clear many times."

"Besides, when I can't explain stuff like this there's this guy that teases me about it."

This guy?

"Okay. Well anyway, permutations are all about thinking in terms of 'for each,' " I said.

3.1.3 An Example

We sat down on a bench. The spring breeze felt nice, making me glad we were outside. I pulled a small notebook out of my pocket, and started my lecture on permutations.

"From the beginning, then. We've got four cards, and we want to think about how many ways there are to arrange them."

"Gotcha."

"By arranging them, I mean we're going to pay attention to the order that we lay them out. For example, we consider this

and this

to be different arrangements."

Yuri nodded. "We pay attention to order. Okay."

"One arrangement like that is called a permutation, a sequence in which order matters."

"A permutation is a sequence where order matters. Got it."

From the look in her eyes, I could tell that Yuri was really following everything. She was an easy student, actually, since she never held back when I lost her attention. She was like Tetra in that respect—neither one pretended to understand something she didn't.

"Another important thing is that you count arrangements with no leaks and no dupes."

"Leaks and dupes? Are those math terms?"

"No, I just mean you can't miss any possible permutations, or your count will be too low, and you can't count the same permutation twice, or your count will be too high. We want exactly the right number."

"Well I'm all for that," Yuri said. "Wouldn't want to be like you, forgetting ties in the dice game."

"Gimme a break, anybody can miscount. That's why we have to use a strategy."

3.1.4 Regularity

"A strategy? For counting?" Yuri said.

"Yep. Specifically, we need to use regularity."

"That makes no sense."

"Well, one way of being sure you count with regularity is to use a tree diagram."

"A diagram? Of trees?"

"A diagram that looks like a tree, like this."

I turned to a blank page in my notebook, and started sketching.

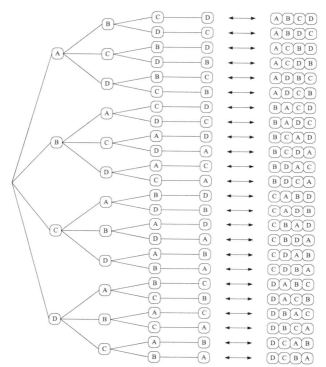

Tree diagram for a permutation.

"See? I've drawn this one to show you how they're related to permutations. Tree diagrams are a helpful way to find regularity."

"Okay, I'll keep that in mind."

The First Divergence

"So let's take a closer look at this diagram," I said. "Starting from the left, there are four divergences. They represent how we can choose any of the four cards as our first selection."

"What's a divergence?"

"A fork, a splitting. Or you can call it branching, in the case of a tree diagram. So in this case, we start with four branches."

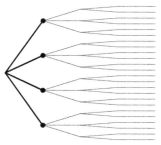

First divergence, into four branches.

The Second Divergence

"On to the next divergence."

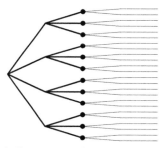

Second divergence, into three branches.

"See how the first four branches split again?"

"Sure."

"Are you starting to see the 'for each' I was talking about?"

"Mmm, maybe."

"This second divergence splits into three branches each. Each of those correspond with the three choices you have for the second card."

"Yeah, I see that. Because we can't reuse the card we used first."

"So now we have three branches *for each* of the first four. Any time you hear 'for each,' that implies multiplication. So the number of branches after the second divergence is $4 \times 3 = 12$."

"Makes total sense."

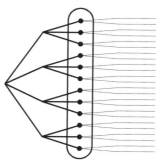

After the second divergence, there are $4 \times 3 = 12$ branches.

The Third Divergence

"Okay, on to the third divergence, but I think you get the picture now. We're just repeating what we did before."

"The twelve branches are going to split into two, right?"

"Yeah, but stick with the tool we're using."

"Tool?"

"The whole 'for each' thing."

"Oh, right. Okay, so *for each* of the twelve branches, there will be two new ones. More multiplication."

"Exactly, and $12 \times 2 = 24$."

"Got it."

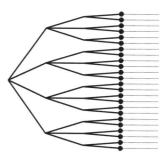

Third divergence: the 4×3 branches each split into two.

The Fourth Divergence

"Fourth and final divergence. There were 24 branches after the third divergence, so—"

Yuri held up a hand.

"Silence! I've got this! The 24 branches branch into... Hang on."

"What's wrong?"

"Can we still say 'branching' if there's just one new branch?"

"Not usually, I guess, but in this case we get a pass. We want to keep using the same terms, after all. That's another form of regularity."

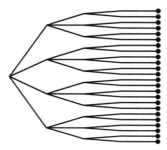

Fourth divergence: the $4 \times 3 \times 2$ branches have one new branch each.

"I'll go with that, then. So for each of the 24 branches, we create one new branch. Multiplication again, but just $24 \times 1 = 24$. So there are 24 branches after the fourth divergence."

"Which means that there are 24 ways to arrange the four cards."

"Okay, I get it."

"Then let's summarize. After each divergence, there's one fewer branches you can choose from, $4 \to 3 \to 2 \to 1$. And we multiplied those numbers in each case."

Start with 4 branches.	←----→		4
Each of those split into 3 branches.	←----→	×	3
Each of those split into 2 branches.	←----→	×	2
Each of those produce 1 branch.	←----→	×	1

"Yeah! So here's that regulating you were talking about!"

Number of permutations of 4 cards $= 4 \times 3 \times 2 \times 1$

"Regularity."

"Regularity, right. Anyway, this regularity definitely makes the counting easier."

"And tree diagrams make regularity easier to find, and bring out the 'for each' multiplication. You get all that, right?"

"Sure. As usual, your explanations are much better than my math teacher's."

"Nah, you're just a better student than you realize."

"Aww, cut it out."

Solution 3-1 (permutations)

The 4 cards

$\boxed{A}, \boxed{B}, \boxed{C}, \boxed{D}$

can be arranged in 24 ways.

"Okay, now that we've found the regularity we were looking for, let's move on."

"To what?"

3.1.5 Generalization

"To generalization," I said.

"Which is?" Yuri asked.

"We found that there are 24 ways to arrange four cards, which is fine, so long as you just have four cards. Generalization means being able to do this when you have five cards, or six, or any number."

Problem 3-2 (permutations)

In how many ways can you arrange n cards?

"How many cards is n?"

"When we generalize, we often use a variable like n to represent numbers. This is called 'generalization through introduction of a variable.' We were able to arrange $\boxed{A}, \boxed{B}, \boxed{C}, \boxed{D}$ in 24 ways specifically because there are four cards. If we can find how many ways we can arrange n cards—in other words, if we can represent the number of permutations using n cards—then we'll know how many arrangements of 5 cards there are, or of 6 cards, or 7 cards... *Any* number of cards."

"Huh."

"This is the difference between mathematics and arithmetic. Remember the first time you saw letters show up in math?"

"Sure, a and b and x and y and all that."

"Right. That was all practice, learning how to use numbers in a more general sense, not only specific numbers like 4."

"Okay, so what does that mean here? For arranging cards?"

"Well, instead of starting with 4 cards, we start with n cards. Then we start multiplying numbers, starting with n and decreasing by one each time."

The first divergence was n branches	←----→	n
Each of those diverged into $n-1$ branches.	←----→	$\times \;(n-1)$
Each of those diverged into $n-2$ branches.	←----→	$\times \;(n-2)$
$\vdots$	$\vdots$	$\vdots$
Each of those diverged into 2 branches.	←----→	$\times \;2$
Each of those diverged into 1 branch.	←----→	$\times \;1$

"I know this!" Yuri said. "That's a factorial, right? You write it as $n!$.

Factorial of n

$$n! = n \times (n-1) \times (n-2) \times \cdots \times 2 \times 1$$

"Good for you," I said.

"Oh, come on. That's not hard."

"So anyway, we've found that there are $n!$ ways of arranging n things."

"And now that I can see this as a tree diagram, I totally get it!"

> **Answer 3-2 (permutations)**
>
> There are $n!$ ways of arranging n cards.

3.1.6 Forging Paths

"What we've done here isn't all that hard," I said. "At least, it isn't hard to just memorize the fact that there are $n!$ ways of arranging n cards. But it's vital to remember how we got there."

"What's getting you all so serious?"

"Because this is important. When you want to consider things mathematically, it's good to start with a specific example, like we did with the four cards."

"You've made that clear."

"But that's not enough. The whole point of examining the specific case is to find the regularity that's hidden within it."

"Sounds like quite a search."

"Miruka calls this 'uncovering its hidden form.'"

Yuri jumped at mention of her idol. "Miruka does?!"

"Tree diagrams are one way of finding regularity. Tables are another. And once you've found it, you can generalize it, usually in the form of an equation or a formula."

"Example to regularity to generalization," Yuri said. "I think I get that... But why? What's the point?"

"Hmm, good question," I said. "I guess the point is moving from knowing by doing to knowing without doing. That's a pretty amazing shift."

Yuri made a face. "Knowing without doing?"

"Sure. Like, knowing you can just calculate $n!$ instead of having to draw a tree diagram for every permutation problem. Applying generalized equations—formulas—makes this possible. That's why formulas are so convenient. But it's hard to see the value in formulas if you haven't done the work with your own hands, and that's why you should never just memorize them. Without truly knowing them, you won't know when and why to apply them."

"Wow, getting kinda heated there."

"Anyway, it's important to consider specifics. There's no getting around that. But it's even more important to follow that up with finding regularity, then generalization. That's true, no matter how small the problem may seem. That's the path you need to forge—specificity to regularity to generality. That path leads to your destination, proof that what you've found is truly correct."

"I'm not sure if I'm ready for that kind of journey quite yet, but if you say so."

"Sure you are. Say you've used your own hands and head to derive this $n!$. Then you will naturally learn that you can always use it to find the number of ways to order a sequence. There's no need for memorization."

"My teacher said we should memorize factorials, though, up to $10!$ at least."

"It can be useful to remember small ones, maybe. I memorized them myself. You never know when you'll run into a 3,628,800 and say, 'Hey, that's a $10!$ there!'"

n	1	2	3	4	5	6	7	8	9	10
$n!$	1	2	6	24	120	720	5040	40,320	362,880	3,628,800

3.1.7 This Guy

"This is why I like it when you teach me things," Yuri said.

She tossed her can into the trash and adjusted her cap. I caught the scent of soap. A new problem came to mind.

"So, Yuri, in how many ways can you arrange the letters in 'animal'?"

"That's easy. $6!$, which is 720. I just memorized that."

"Not quite."

"Huh? Why not?"

"Because there are two a's in 'animal.' So you can't use the $n!$ formula."

"So you've tricked me twice—by picking a word with a repeated letter, and by making it 'aminal.' No, 'animal.' So I guess the answer is—"

Yuri gave me a sharp poke in the ribs. "Ack!"

"—factorial-ack," she finished.

"That kinda hurt, Yuri."

"Oh, gimme a break. Just a little sign of affection."

"A painful one. By the way, what brought up all this stuff anyway?"

"Well, there's this guy in my class who's good at math and just won't let me forget it. He's always posing these tricky problems, and one day he asked me if I knew about permutations. When I couldn't explain what they're all about, it just... I don't know, it made me mad."

"So you're learning because you don't want 'this guy' to show you up?"

"Sure! Next time I see him, at least. He said he has to take some time off from school for some reason."

Problem 3-3 (sequences with repeating elements)

In how many ways can you arrange the six letters in 'animal'?

3.2 COMBINATIONS

3.2.1 In the Library

The silence in the library was shattered by a shouted "Whoooa!" that could only have come from Tetra. I turned just in time to see her falling in a cloud of cards.

I was in the library after school, meditating on how quickly my sixteenth year (2^{2^2}!) had passed by, bringing me to a prime seventeen. The days flowed by as always, taking me closer to my college entrance exams. Day after day of classes, followed by study in the library. The same pace as always, with just one difference—I seemed to be getting busier with other things, which left less time for math.

"Ouch..." Tetra said as she gathered up cards. I left my seat to help her.

"You okay?" I said.

I picked up a card, and saw it was from Mr. Muraki.

3.2.2 Permutations

"Huh, this one's about permutations," I said, handing the card back to Tetra. "What a coincidence—I was just teaching Yuri about these."

Permutations formula

The number of possible ways of selecting k elements in order from among n elements is

$$_nP_k = \frac{n!}{(n-k)!}.$$

"I didn't go this far with Yuri, though, generalizing how you can select just k elements. We used all n."

"Which would be, um, $_nP_n$."

"Right."

$$\begin{aligned}
_nP_n &= \frac{n!}{(n-n)!} && \text{n = k, from the definition of } _nP_k \\
&= \frac{n!}{0!} && \text{because } n - n = 0 \\
&= \frac{n!}{1} && \text{because } 0! = 1 \\
&= n!
\end{aligned}$$

"About this permutations formula $_nP_k$, though..." Tetra said.

$$_nP_k = \frac{n!}{(n-k)!}$$

"What about it?"

"We should always get an integer answer, so it seems kind of strange to have a fraction in the definition. That doesn't quite sit well with me."

"It's a fraction in form, but simplifying will always end up giving us an integer."

"I guess that's the strange part."

"Well, let's take an example and see if that convinces you. Say you wanted to select two items out of five."

- There are five ways to make the first selection.
- There are four ways to make a second selection after each first selection.

"In other words, the permutation $_5P_2$ becomes this."

$$_5P_2 = \text{number of ways of selecting 2 items from 5} = 5 \times 4$$

"Sure, I see that. Multiplication of descending integers, 5 and 4."

"Right. We can write multiplication of descending integers using the factorial n!."

"Whoa, hang on. The factorial of 5 is different from $_5P_2$. The factorial has that extra tail. Like this."

$$5! = \underbrace{5 \times 4}_{_5P_2} \times \underbrace{3 \times 2 \times 1}_{\text{tail}}$$

"But among the $5 \times 4 \times 3 \times 2 \times 1$, all we need for $_5P_2$ is the 5×4 part, right? So we can just get rid of the $3 \times 2 \times 1$, by dividing it out. Which, it turns out, is 3!. So we can write $_5P_2$ using just factorials."

$$_5P_2 = 5 \times 4$$

$$= \frac{5 \times 4 \times \overbrace{3 \times 2 \times 1}^{\text{tail}}}{\underbrace{3 \times 2 \times 1}_{\text{tail}}}$$

$$= \frac{5!}{3!}$$

"Yeah, I see!" Tetra said, nodding. "You've chopped the tail off!"

"That's how we do it with the specific values 5 and 2, at least. We can use variables n and k to get the equation for the permutation $_nP_k$."

$$_nP_k = n \times (n-1) \times (n-2) \times \cdots \times (n-k+1)$$

$$= \frac{n \times (n-1) \times (n-2) \times \cdots \times (n-k+1) \times \overbrace{(n-k) \times \cdots \times 2 \times 1}^{\text{tail}}}{\underbrace{(n-k) \times \cdots \times 2 \times 1}_{\text{tail}}}$$

$$= \frac{n!}{(n-k)!}$$

"Oh! The denominator $(n-k)!$ is the tail!"

3.2.3 Combinations

"So if order is important we call it a permutation," Tetra said, "and if we don't care what order the elements come in, it's a combination, right?"

"Right. For combinations we use the notation $\binom{n}{k}$ or $_nC_k$."

Permutation	$_nP_k$	Number of ways of selecting k items from among n, paying attention to order.
Combination	$_nC_k$ or $\binom{n}{k}$	Number of ways of selecting k items from among n, when order doesn't matter.

"As an example, let's say we want to choose two of five cards: A, B, C, D, E. If we need to pay attention to what order the cards are chosen in—a permutation—then there are $_5P_2 = 5 \times 4 = 20$ possibilities."

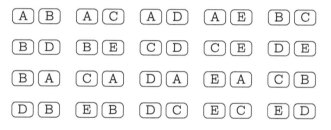

Selecting 2 cards from among 5, paying attention to order (permutation).

"Right," Tetra said.

"If we do the same thing, but don't care about what order we draw the cards in, there's only 10 patterns."

Selecting 2 cards from among 5, not paying attention to order (combination).

"Writing the combinations as $\binom{5}{2}$, we get this."

$$\binom{5}{2} = \frac{5!}{2!\,3!} = 10$$

"They have different numbers because, for example, as a permutation ⒜⒝ and ⒝⒜ are considered different things, but as a combination they're the same thing."

"Right," Tetra said. "Combinations combine the similar pairs."

"So how many pairs are similar?"

"Well, it looks like there are twice as many permutations as there are combinations."

"Right. Thinking carefully about it, *permutations* of our selections are duplicates. Like, when we're selecting two cards, ⒜⒝ and ⒝⒜ are duplicates. This is a permutation of two selected cards, or $_2P_2 = 2!$."

"Huh."

"First we make our selections considering order. But when we do that, it causes duplications. So we can divide out the duplications to find the number of combinations."

$$\begin{aligned}
&\text{Ways of selecting 2 cards from 5, not considering order} \\
&= \frac{\text{Ways of selecting 2 cards from 5, considering order}}{\text{Ways of selecting 2 cards, considering order}} \\
&= \frac{_5P_2}{_2P_2} \\
&= \frac{5 \times 4}{2 \times 1} \\
&= 10
\end{aligned}$$

"Make selections considering order, then divide out the duplicates. Got it."

"Once we look at things this way, generalization is easy."

$$\begin{aligned}&\text{Ways of selecting k cards from n, not considering order}\\&=\frac{\text{Ways of selecting k cards from n, considering order}}{\text{Ways of selecting k cards, considering order}}\\&=\frac{{}_nP_k}{{}_kP_k}\\&=\frac{\frac{n!}{(n-k)!}}{k!}\\&=\frac{n!}{k!\,(n-k)!}\end{aligned}$$

"Cool!"

"This fraction $\frac{n!}{k!(n-k)!}$ has shown up again, but remember that we're just dividing out duplicates, so the result will always be an integer."

"Okay, that pretty much convinces me, now that I see what the n! and the k! and the (n − k)! mean."

Definition of combinations

The number of ways of selecting k items from among n without considering order is

$$_nC_k = \binom{n}{k} = \frac{n!}{k!\,(n-k)!}.$$

3.2.4 Animal

Tetra collected the cards that were scattered on the desk, then put some finishing touches on her notes. Flighty though she could be, she was very good about taking notes when learning something new. Maybe it had something to do with her love of words. Which reminded me...

"So, Tetra."

"Yeah?" she said, looking up from her notes.

"How many ways can you arrange the six letters in 'animal'?"

"Let's see. We need to pay attention to order, so that's a permutation. So I guess the answer is 6!... No no no, wait! There are two a's, aren't there. So hang on. There's six letters in all, so we get this."

$$\text{Permutations of 6 letters} = 6!$$

"But that includes duplicates. We don't differentiate between the two a's, so we have to divide by a permutation of two letters."

$$\text{Permutations of 2 letters} = 2!$$

"So it should be this."

$$\begin{aligned}\text{Ways of arranging the letters in ``animal''} &= \frac{\text{Permutations of 6 letters}}{\text{Permutations of 2 letters}} \\ &= \frac{6!}{2!} \\ &= 6 \times 5 \times 4 \times 3 \\ &= 360\end{aligned}$$

"Exactly right," I said. "It's the same thing in the end, but you can also differentiate between the two a's like this."

$$\begin{aligned}\text{Arrangements of letters in animal} &= \frac{\text{Permutations of } \overset{1}{a} \cdot n \cdot i \cdot m \cdot \overset{2}{a} \cdot l}{\text{Permutations of } \overset{1}{a} \cdot \overset{2}{a}} \\ &= \frac{6!}{2!} \\ &= 6 \times 5 \times 4 \times 3 \\ &= 360\end{aligned}$$

"Wow, 360 arrangements?" Tetra said. "Let's see... 'animal,' 'animla,' 'anilma,' 'anlima'..."

"Please don't tell me you're going to try to say every one."

Answer 3-3 Permutations (sequences with repeating elements)

The letters in 'animal' can be arranged in 360 ways.

3.2.5 The Binomial Theorem

I pulled another card out of the stack.

"Oh look, the binomial theorem," I said. "The most famous theorem related to combinations."

The binomial theorem

$$(a+b)^n = \sum_{k=0}^{n} \binom{n}{k} a^{n-k} b^k$$

"Sounds familiar," Tetra said. "Didn't you tell me about this before? But there's so many letters..."

Tetra made a face showing what she thought about lots of letters.

"Think of it as an opportunity for practicing how to put things in concrete form," I said.

"Meaning?"

"Well, there's lots of letters—er, variables—in formulas like this because they're highly generalized. It's like you're seeing the results after somebody has done a lot of 'generalization through the introduction of a variable' work for you. Then you can go backwards, plugging specific values into the variables and seeing how things fall out. It's very satisfying to confirm that the formula works."

"That's kind of like the opposite of 'generalization through the introduction of a variable,' isn't it."

"Yeah, sure. 'Particularization through substitution for variables,' I guess. For example, say that in the binomial theorem we have $n = 1$."

The Solitude of 17,179,869,184

$$(a+b)^1 = \sum_{k=0}^{1} \binom{1}{k} a^{1-k} b^k \qquad \text{Letting } n = 1 \text{ in the B.T.}$$

$$= \underbrace{\binom{1}{0} a^{1-0} b^0}_{\text{for } k=0} + \underbrace{\binom{1}{1} a^{1-1} b^1}_{\text{for } k=1} \qquad \text{without the } \sum \text{ symbol}$$

$$= 1a^{1-0} b^0 + 1a^{1-1} b^1 \qquad \text{using } \binom{1}{0} = 1, \binom{1}{1} = 1$$

$$= 1a^1 b^0 + 1a^0 b^1$$

$$= a^1 + b^1$$

$$= a + b$$

"Looks like it worked," Tetra said, "since $(a+b)^1$ does equal $a+b$."

"Sure. Let's do it again, with $n=2$."

$$(a+b)^2 = \sum_{k=0}^{2} \binom{2}{k} a^{2-k} b^k$$

$$= \underbrace{\binom{2}{0} a^{2-0} b^0}_{\text{for } k=0} + \underbrace{\binom{2}{1} a^{2-1} b^1}_{\text{for } k=1} + \underbrace{\binom{2}{2} a^{2-2} b^2}_{\text{for } k=2}$$

$$= 1a^{2-0} b^0 + 2a^{2-1} b^1 + 1a^{2-2} b^2$$

$$= 1a^2 b^0 + 2a^1 b^1 + 1a^0 b^2$$

$$= a^2 + 2ab + b^2$$

"This looks familiar too!"
"Once more, with $n=3$."

$$(a+b)^3 = \sum_{k=0}^{3} \binom{3}{k} a^{3-k} b^k$$

$$= \underbrace{\binom{3}{0} a^{3-0} b^0}_{\text{for } k=0} + \underbrace{\binom{3}{1} a^{3-1} b^1}_{\text{for } k=1} + \underbrace{\binom{3}{2} a^{3-2} b^2}_{\text{for } k=2} + \underbrace{\binom{3}{3} a^{3-3} b^3}_{\text{for } k=3}$$

$$= 1a^{3-0} b^0 + 3a^{3-1} b^1 + 3a^{3-2} b^2 + 1a^{3-3} b^3$$

$$= 1a^3 b^0 + 3a^2 b^1 + 3a^1 b^2 + 1a^0 b^3$$

$$= a^3 + 3a^2 b + 3ab^2 + b^3$$

"Oh, I get it! The binomial theorem is a generalization of how you expand things like $(a+b)^2$ and $(a+b)^3$. I'm a little embarrassed it took me so long to realize that. Sorry for being so slow."

I chuckled. "You aren't slow. There are lots of times when concrete numbers show you things the letters didn't. That's natural."

"Well that's a relief," Tetra said with the cutest of smiles.

"Remember back in junior high when you had to memorize things like $(a+b)^2 = a^2 + 2ab + b^2$, expansions into a sum of products? Well, it's probably worth memorizing ones you use a lot, but it's best if you understand them using the binomial theorem."

"You bet!"

"Do you see how each term in the binomial theorem contains an $\binom{n}{k}$ combination?"

"I do! You told me about this before! Like, $(a+b)^3$ is a product of three factors, right?"

$$(a+b)^3 = \underbrace{(a+b)}_{\text{factor 1}}\underbrace{(a+b)}_{\text{factor 2}}\underbrace{(a+b)}_{\text{factor 3}}$$

"When we expand this, we select an a or a b from among each of those three factors and multiply them."

$(\boxed{a}+b)(\boxed{a}+b)(\boxed{a}+b)$	$\to$	$aaa = a^3b^0$
$(\boxed{a}+b)(\boxed{a}+b)(a+\boxed{b})$	$\to$	$aab = a^2b^1$
$(\boxed{a}+b)(a+\boxed{b})(\boxed{a}+b)$	$\to$	$aba = a^2b^1$
$(\boxed{a}+b)(a+\boxed{b})(a+\boxed{b})$	$\to$	$abb = a^1b^2$
$(a+\boxed{b})(\boxed{a}+b)(\boxed{a}+b)$	$\to$	$baa = a^2b^1$
$(a+\boxed{b})(\boxed{a}+b)(a+\boxed{b})$	$\to$	$bab = a^1b^2$
$(a+\boxed{b})(a+\boxed{b})(\boxed{a}+b)$	$\to$	$bba = a^1b^2$
$(a+\boxed{b})(a+\boxed{b})(a+\boxed{b})$	$\to$	$bbb = a^0b^3$

"Then we add together the resulting eight multiplications, $aaa, aab, aba, \ldots, bbb$. We use coefficients to show how many similar terms popped up, so that's the number of those combinations. For instance, the coefficient on a^2b^1 is the number of ways of selecting two a's from among the three factors, the number of two-a combinations."

"Exactly," I said. "The coefficients are very interesting."
I wrote out some expansions, circling the coefficients.

$$(a+b)^0 = \boxed{1}$$
$$(a+b)^1 = \boxed{1}a + \boxed{1}b$$
$$(a+b)^2 = \boxed{1}a^2 + \boxed{2}ab + \boxed{1}b^2$$
$$(a+b)^3 = \boxed{1}a^3 + \boxed{3}a^2b + \boxed{3}ab^2 + \boxed{1}b^3$$

"Interesting how?"

"Have you ever heard of—"

Then it hit me, an aroma of citrus. I spun around to see a bright smile.

"Pascal's triangle, right?" Miruka said.

3.3 Allocations of 2^n

3.3.1 Pascal's Triangle

"Don't let me interrupt," Miruka said.

"I've read something about Pascal's triangle somewhere..." Tetra said. "Something about adding up adjacent numbers?"

"Right," I said. "You start with a 1 ..."

I wrote a Pascal's triangle in my notebook, something I had done many times in the past. There was something very enjoyable about writing them out by hand.

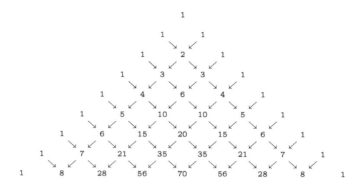

Pascal's triangle.

Tetra watched from beside me and Miruka from behind as I wrote. The pressure was almost too much.

"Well? Go on?" Miruka said.

"Right. Uh, the coefficients that appear in an expansion of $(a + b)^n$ can also be found in the nth row of Pascal's triangle. Taking the topmost row as row 0—"

0th row $\quad (a+b)^0 \qquad\qquad\qquad = ①$

1st row $\quad (a+b)^1 \qquad\qquad = ①a \quad + \quad ①b$

2nd row $\quad (a+b)^2 \qquad = ①a^2 \quad + \quad ②ab \quad + \quad ①b^2$

3rd row $\quad (a+b)^3 = ①a^3 \quad + \quad ③a^2b \quad + \quad ③ab^2 \quad + \quad ①b^3$

"Hey, cool! But hang on, there's something strange here. We create Pascal's triangle by adding numbers, so doesn't it seem kind of weird that it gives the same numbers we got from multiplications when creating combinations?"

"Huh, never really thought of it that way."

"The diagram confuses things," Miruka said, pushing her glasses up her nose. "People tend to want to draw the numbers with symmetry, because of the name, but Pascal's triangle is best viewed as a table."

		k							
		0	1	2	3	4	5	6	7	8
	0	$\binom{0}{0}$								
	1	$\binom{1}{0}$	$\binom{1}{1}$							
	2	$\binom{2}{0}$	$\binom{2}{1}$	$\binom{2}{2}$						
	3	$\binom{3}{0}$	$\binom{3}{1}$	$\binom{3}{2}$	$\binom{3}{3}$					
n	4	$\binom{4}{0}$	$\binom{4}{1}$	$\binom{4}{2}$	$\binom{4}{3}$	$\binom{4}{4}$				
	5	$\binom{5}{0}$	$\binom{5}{1}$	$\binom{5}{2}$	$\binom{5}{3}$	$\binom{5}{4}$	$\binom{5}{5}$			
	6	$\binom{6}{0}$	$\binom{6}{1}$	$\binom{6}{2}$	$\binom{6}{3}$	$\binom{6}{4}$	$\binom{6}{5}$	$\binom{6}{6}$		
	7	$\binom{7}{0}$	$\binom{7}{1}$	$\binom{7}{2}$	$\binom{7}{3}$	$\binom{7}{4}$	$\binom{7}{5}$	$\binom{7}{6}$	$\binom{7}{7}$	
	8	$\binom{8}{0}$	$\binom{8}{1}$	$\binom{8}{2}$	$\binom{8}{3}$	$\binom{8}{4}$	$\binom{8}{5}$	$\binom{8}{6}$	$\binom{8}{7}$	$\binom{8}{8}$

Combinations $\binom{n}{k}$.

		k								
		0	1	2	3	4	5	6	7	8
	0	1								
	1	1	1							
	2	1	2	1						
	3	1	3	3	1					
n	4	1	4	6	4	1				
	5	1	5	10	10	5	1			
	6	1	6	15	20	15	6	1		
	7	1	7	21	35	35	21	7	1	
	8	1	8	28	56	70	56	28	8	1

Combinations $\binom{n}{k}$ (values).

"Like these tables show," Miruka said. "Pascal's triangle is created from these calculations."

$$\binom{n-1}{k-1} \quad \binom{n-1}{k}$$
$$\searrow \quad \downarrow$$
$$\binom{n}{k}$$

"In other words, we're applying this recurrence relation for integers n and k that satisfy $0 < k \leqslant n$."

$$\binom{n}{k} = \binom{n-1}{k-1} + \binom{n-1}{k}$$

"Lots of n's and k's in there," Tetra said. "What does this mean, exactly?"

Miruka just pointed at me, a silent demand that I do the explaining.

"Well, we can read the recurrence relation like this."

The number of ways of selecting k items from among n
= The number of ways of selecting k − 1 items from among n − 1
+ The number of ways of selecting k items from among n − 1

"Oh, okay. So..."

"This recurrence relation shows a separation of cases," I said.

"How so?"

"Let's give it a try, using $n = 4$ and $k = 2$. Say you have four cards—⟨A⟩, ⟨B⟩, ⟨C⟩, and ⟨D⟩—and you're selecting two of them. We're looking for cases where, for example, we select an ⟨A⟩ card."

"So were separating into cases where we choose an ⟨A⟩, and cases where we don't?"

"Right. In cases where we select an ⟨A⟩, we can consider the number of combinations as those combinations where we've also selected one of the three non-⟨A⟩ cards. In other words, combinations of selecting k − 1 cards from among n − 1 cards."

"So we're choosing two cards, but since we've already chosen an ⟨A⟩, it's just a question of what the final selection is."

"That's right. And in cases where we *didn't* choose an ⟨A⟩, the number of combinations is those combinations where we choose two cards from among the three non-(A) cards. In other words, k cards from among n − 1 cards."

"And that's the second part of the sum."

I nodded. "That sum shows the two cases that together show combinations of selecting two cards from among four."

$$\binom{n}{k}$$ Combinations of choosing k cards from among n

$$= \binom{n-1}{k-1}$$ Combinations of choosing k − 1 non-(A) cards from among n − 1 cards

$$+ \binom{n-1}{k}$$ Combinations of choosing k cards from among the n − 1 non-(A) cards

"Now that I see it like this, the equation makes perfect sense!"

3.3.2 Bit Patterns

Miruka stood and walked to a window, observing the sycamore on the other side. After a pause she spun back toward us, fanning her hair.

"Let's talk about bit patterns," she said. "Using n bits, we can represent 2^n n-digit binary strings."

"Can you show me an example of that?" Tetra asked.

"Sure. Say that $n = 5$. Then we can create 5-digit binary strings, 00000 through 11111. There's $2^5 = 32$ of those."

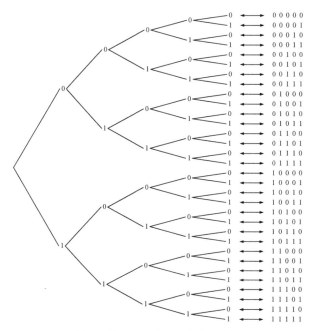

Tree diagram for a bit pattern

"Now let's sort the bit patterns by 1-count. For instance, 00000 doesn't have any ones in it, so its 1-count is 0. Similarly, 00101 has a 1-count of 2, 10110 has a 1-count of 3, and so on. Of course the largest 1-count will be 5, for 11111. This histogram sums up the whole bunch."

				11000	11100	
				10100	11010	
				10010	10110	
				10001	01110	
				01100	11001	
		10000	01010	10101	11110	
		01000	01001	01101	11101	
		00100	00110	10011	11011	
		00010	00101	01011	10111	
	00000	00001	00011	00111	01111	11111
Number of ones	0	1	2	3	4	5
Number of patterns	1	5	10	10	5	1

Distribution of 1-counts among bit patterns.

"The numbers of patterns at the bottom are like numbers of combinations, right?" Tetra asked.

"Very good," Miruka replied. "It is, because saying 'k bits among n are 1' is the same as saying 'choosing k items from among n.' You're just setting bits to 1 instead of drawing cards."

3.3.3 Exponential Explosions

"Pretty cool that you can represent 32 numbers using just 5 bits," Tetra said.

"If you think 32 from 5 is cool, I guess you've never experienced the awesomeness of exponential explosions," Miruka said.

"Uh, I guess not?"

"With one bit, you can assign numbers to two people—0 or 1."

"Yeah, sure," Tetra said, eyeing her suspiciously. "So?"

"With two bits you can number four people—00, 01, 10, and 11."

"Yeah...?"

"And so on and so on. If you're using n bits, you can assign numbers to 2^n people, from $\underbrace{000\cdots 0}_{n \text{ bits}}$ to $\underbrace{111\cdots 1}_{n \text{ bits}}$. You're with me so far, right?"

"Sure, n bits give 2^n binary numbers."

"So tell me," Miruka said. "How many bits would you need to assign every person in the world a unique number? For simplicity, let's say that's 10 billion people."

"Seriously? A unique number for ten billion people?"

"Yep."

"Wow, I don't know. Like, 10,000 bits?"

"Way less than that."

"Seriously? Then, around 3000 maybe?"

"Still way too many."

"300?"

"Try 34."

"No way!" Tetra said. "34 bits can seriously generate ten billion bit patterns?"

"A lot more than that, actually, but 33 bits doesn't quite cut it."

$$2^{33} = 8{,}589{,}934{,}592$$
$$2^{34} = 17{,}179{,}869{,}184$$

$$2^{33} < 10 \text{ billion} < 2^{34}$$

"I can't believe you don't need at least 300," Tetra said.

"I've heard that the volume of the entire universe is around 2^{280} cubic centimeters," Miruka said, "so 300 would be way overkill. I mean, think about it. With just 280 bits—" Miruka placed her hands on Tetra's shoulders, and moved her face in so close that their noses and lips were nearly touching. "—you could divide the entire universe into one-centimeter cubes and number each one."

"Uh, er, uh..." Tetra stammered, nearly paralyzed.

3.4 THE SOLITUDE OF POWERS

3.4.1 On the Road Home

I walked home alone, daydreaming about tree diagrams.

Just 34 bifurcations would end up as 17,179,869,184 branches, more than enough to assign one to every human being on earth. 34 decisions produce 17,179,869,184 possible results. Considering all the branches I've taken in my past, how many possible "me"s are their that I am not?

3.4.2 At Home

"I'm home!" I called out.

My mother approached me, whispering for some reason.

"Yuri's here!"

"Yeah?" I whispered back, though unsure why.

I entered the room, and saw Yuri slumped in a chair. She was looking down, hands deep in the pockets of her pullover. Even her ponytail had lost its normal bounce.

"You okay?" I asked.

"Not at all," she said, tears brimming in her eyes. "He's... He's going to change schools. I don't know what to do."

Yuri closed her eyes, folded forward, and covered her face with her hands.

She could only have been talking about that guy in her class, the one who she did math problems with after school. As much as she complained about him, he was someone to talk to, someone to learn alongside, someone to have fights and make up with. He was moving far away or something, and now she had lost all that.

The difference of just a single bit can change everything, and every day is full of new branches. Life is a journey through an infinitude of divergences.

Yuri was still sitting with her hands covering her face, her shoulders slightly trembling. I was unsure what I should say, but I knew how she felt. I moved closer, and put my hand on her head.

Number of atoms in the planet	2^{170}
Number of atoms in the galaxy	2^{223}
Volume of the universe	2^{280} cm^3

BRUCE SCHNEIER
Applied Cryptography

CHAPTER 4

Uncertain Certainty

> Oh that there had been but one or two, nay, or but one soul saved out of this ship, to have escaped to me, that I might but have had one companion, one fellow-creature, to have spoken to me and to have conversed with!
>
> DANIEL DEFOE
> *Robinson Crusoe*

4.1 CERTAIN CERTAINTY

4.1.1 What Division Means

The following Saturday, Yuri was back in my room, browsing through my books as usual. I was flipping through a stack of vocabulary cards, answering the odd question she threw my way. Her tears from the other day had dried up, and she seemed to have largely reverted to her old self. I couldn't be sure what was going on beneath her normal-seeming chirpy exterior, though.

"Tell me what division means," she said, out of the blue.

"Such a philisophical question."

"Remember when I asked you about that dice game? The one you got *wrong*?"

Like you would ever let me forget it...

Alice and Bob are playing a dice game: each throws a single die, and whoever gets the larger number wins. What is the probability that Alice will win?

"I remember."

"When we were talking about that, you wrote an equation like this."

Yuri found some paper and a pencil on my desk, and wrote

$$\text{Probability of Alice winning} = \frac{\text{Alice wins (15)}}{\text{All cases (36)}} = \cdots = \frac{5}{12}.$$

"I did."

"Well that's division, right? Since it's a fraction."

"Sure, writing $\frac{\text{numerator}}{\text{denominator}}$ is the same as writing numerator ÷ denominator, so you can either say that Alice's chances of winning are $\frac{5}{12}$ or that they're $5 \div 12 = 0.4166\cdots$. Same thing."

"But why are probabilities division problems?"

"Huh? You mean you want to know why we divide the number of cases we're interested in by all possible cases?"

"Well, I kind of see why we do that, but I kind of don't. I want to get rid of that 'kind of don't' part."

"Okay, I'm not quite clear what you don't understand, but let's try to clear things up."

I took the paper and pencil, and Yuri pulled a chair up next to me.

"Are you wearing lipstick?" I asked.

"What, this?" she said, pointing to her mouth. "It's just lip gloss," she said, smirking.

"Oh, right. Anyway, so Alice throws a die, and gets one of six results: ⚀, ⚁, ⚂, ⚃, ⚄, or ⚅. Then Bob does the same, and gets one of the same results."

"Yep, that's the game."

"So for each of Alice's six possibile outcomes, there are six possible outcomes for Bob. And that 'for each of' implies—"

"Multiplication!"

"You got it. So in all there are 6 cases × 6 cases = 36 cases, and each of those cases has an equal probability of occuring. Alice wins 15 of the 36 cases, so her probability of winning is $\frac{15}{36}$. Simplify, and you get $\frac{5}{12}$. So what about that don't you get?"

"When you describe it like that, it comes so close to making sense. But I still don't see where the division comes in."

"Hmm, okay. Let's try to look at it another way."

I tried to imagine how Yuri's mind was working, what wasn't making sense, so that I could come up with a better description. There was no point in getting frustrated, that wouldn't help her learn anything. The better approach was to try to see things from her perspective, and change my presentation to better fit her understanding.

"How about this," I said. "You know how division considers the whole as 1?"

"Let's pretend I don't."

"It's like, if you took the entirety of some thing to be 1, then division tells you what part of that whole you're interested in. You call that a proportion, or a ratio."

"Still not making a whole lot of sense."

"Okay, say you have a candy bar that's 13 cm long, and you eat 6.5 cm of it. What proportion of the candy bar have you eaten?"

"That's easy. Half of it."

"Right, the whole length is 13 cm, and the 6.5 cm length that you've eaten is half, or $\frac{1}{2}$."

$$\text{Proportion of eaten length} = \frac{\text{Eaten length (6.5 cm)}}{\text{Total length (13 cm)}} = \frac{6.5}{13} = \frac{1}{2}$$

"I graduated elementary school, you know."

"Bear with me. So if you take the total length to be 1, the length you've eaten is $\frac{1}{2}$."

I summarized this in the notebook.

> Letting the <u>total amount</u> be 1,
>
> what is the proportion of the <u>amount of interest</u>?

"The answer to this question is a fraction, in other words a calculation by division."

"I get all that. But a probability isn't a length!"

"No, it's not, but it's similar to a *proportion* of length. Or a proportion of area, or volume, or anything, really. Like this."

> Letting <u>what definitely happens</u> be 1,
>
> what is the proportion of the <u>cases of interest</u>?

"We call the answer to that a probability."

"Okay, that's clearing things up a little. You said a probability is like a proportion of area, right? That made me think of a table."

Yuri took the paper and started drawing.

		Bob					
		⚀	⚁	⚂	⚃	⚄	⚅
	⚀	Tie	Bob	Bob	Bob	Bob	Bob
	⚁	Alice	Tie	Bob	Bob	Bob	Bob
Alice	⚂	Alice	Alice	Tie	Bob	Bob	Bob
	⚃	Alice	Alice	Alice	Tie	Bob	Bob
	⚄	Alice	Alice	Alice	Alice	Tie	Bob
	⚅	Alice	Alice	Alice	Alice	Alice	Tie

"We can make this look like a square by showing Alice wins as 'A,' Bob wins as 'B,' and ties as 'T.'"

T	B	B	B	B	B
A	T	B	B	B	B
A	A	T	B	B	B
A	A	A	T	B	B
A	A	A	A	T	B
A	A	A	A	A	T

"Interesting," I said, scanning Yuri's table.

"So if we say that the area of the entire square is 1, then asking 'what's the probability that Alice wins' is the same as asking 'what's the area of the 'A' squares,' right?"

"Exactly right! I couldn't have said it better myself."

"Keep it coming. More praise."

"That's my daily quota."

"Cheapskate. You need to increase your proportion of kind words." Yuri stuck her tongue out at me. "Anyway, we aren't really finding an area, we're just counting. There's 36 squares, and 15 marked 'A.'"

"Right. Also important is that each of the 36 squares occurs with the same certainty."

Yuri stared at the square, thinking about something. She looked up suddenly, her ponytail glinting gold.

"Hey! Doesn't this mean we don't have to throw 2 dice? We can just spin a 36-slotted roulette wheel!"

"A what?"

"Like this!" she said, and started drawing again.

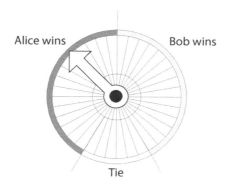

A 36-slotted roulette wheel that duplicates the dice game.

"Yeah, that works. There's 15 results where Alice wins, 15 where Bob wins, and 6 ties."

"If none of those are wins, then the probability of a win is 0."

"Because no winning segments means a win can never occur. And if all of the segments show a win for Alice—"

"—then the probability of an Alice win is 1. That's the only thing that *can* happen."

"Right. All events lay somewhere between 'never happens' and 'always happens.' As an inequality, it looks like this."

$$0 \leqslant \text{probability} \leqslant 1$$

"Probabilities can never be negative," I continued, "and never greater than 1. They're always 0% or greater, and 100% or less."

Yuri laughed at that.

"I knew it! There's no such thing as a 120% probability!"

"What are you talking about?"

"The other day, my teacher said something had a 120% probability. He got mad at me when I said that sounded wrong, mathematically speaking."

"Making fun of your teachers is a losing strategy," I said. "I'm pretty sure he was just exaggerating."

"Yeah, yeah," Yuri said. "But what was really funny was after class, when that guy said—"

Yuri's words suddenly cut off. I looked up at her, but she spun around and reached for something on my bookshelf.

"What's wrong?" I said, but it was a while before she replied.

"Let me borrow this book," she said, and carried it off to the living room.

4.2　Uncertain Certainty

4.2.1　Same Certainties

"Ugh, I hate probability," Tetra said over her bento box.

We were eating our lunches on the school roof. It was a cloudy day, but warm nonetheless.

"Yeah? Why?" I said through a mouthful of the roll I had bought at the school store.

"I get the thing about treating 'the whole' as 1, but I'm still not clear on the 'same certainty' part." Tetra poked at her omelet for a minute before continuing. "Actually, I guess it's the word 'certainty' that causes me problems. Maybe I should stick with 'probability,' in the sense of how probable something is."

Typical of Tetra, I thought. *Getting hung up on the fine nuances of words.*

"Anyway," she continued, "In these problems we talk about how each face on a die can come up with the same probability, but how can we say that with... uh, certainty."

"With a probability of $\frac{1}{6}$ for each face, you mean."

"Yeah, I see how if each face comes up with the same probability, then that probability is $\frac{1}{6}$, but what lets us say the first part? That every face definitely comes up with the same certainty?"

Tetra put down her chopsticks and looked at me. I hesitated, because I finally understood just what she was asking.

"Well, if the die isn't deformed, then each face should have the same shape, so—"

"But when you start talking about the size and shape of dice, you aren't talking math anymore."

"Yeah, I guess now we're in the realm of physics, or engineering maybe."

"As you can tell, I'm having problems saying exactly what it is that's bothering me."

Tetra popped the last of her omelet in her mouth and chewed.

4.2.2 The True Weapon

I shoved the bun wrapper in my pocket, and thought about what Tetra has said. Her concern boiled down to something like this:

> What does it mean to say that each face of a die comes up with equal certainty?

It was an honest question, but one I was unsure how to answer. One that I had never even considered, really. Just being able to calculate permutations and combinations had carried me through all the probability problems I'd ever done. But Tetra saw things differently. She retained her doubts until she *knew* that she knew something.

I noticed something about her: I had always considered her unique perspectives to be her keenest weapon, while she thought that her persistence was. But her true strength was neither of those—her most powerful weapon was her ability to know what she didn't know.

When we'd first met the year before, there were many things she didn't know. Through our conversations she had learned about primes, absolute values, and summation. She had learned about the importance of definitions, and the importance of equations. But I suspected that the more she learned, the deeper her worries of lacking total comprehension became.

I asked her if that was the case, and she waved her hands in negation.

"No, no. I'm just slow. I wish I wasn't, but it just takes me forever to learn new things."

"It's not just you," I said. "The problems that real mathematicians face aren't ones that can be solved right away, so I think it's really important to be able to keep poking at something that's nagging at you until you really get it."

"Well, thanks for that—it means a lot. Maybe my slowness isn't totally a bad thing after all."

"There's nothing bad about you."

"Thanks. I'll keep at it. There's still a lot I want to learn. You and Miruka have really inspired me, you know. You make me feel like I'm not alone."

"Not alone?"

"Yeah. When I can't solve a math problem, I get really worked up, my hands start to sweat... But when that happens, I take a deep breath and think about all the encouragement you guys have given me. That lets me relax and get back to work in a new frame of mind. I remember once you told me that when we're working on math problems, we're like budding mathematicians."

"I did."

"I remember that, and tell myself not to worry that a problem is taking too much time, just to tackle what's in front of me without freaking out. That everyone does this. I might be working on the problem alone, but I'm not really alone. You're there, supporting me from somewhere inside. So no, I'm not alone, even if we're working on different problems. By thinking through things by myself, by fighting through the rough spots on my own, I grow closer to you. You're the one who introduced me to this world, after all."

Tetra looked away, trying to hide a blush.

"That's why... Um, I wanted to tell you, that I, uh—"

Just then, I felt a drop of something cold and wet on my face.

"Uh oh, it's starting to rain," I said. The raindrops started coming faster and fatter.

"My lunch!" Tetra squealed.

We scrambled to gather our things, and hurried inside. As we headed down the stairs, Tetra turned and said, "Thank you for everything!"

4.3 An Experiment in Certainty

4.3.1 Interpreter

After classes were over, I went to the library as usual. I noticed that new leaves were budding on the sycamore outside the window, a bright contrast to the blue skies beyond; the clouds that had brought the sudden squall had vanished.

When I arrived, Tetra and Lisa were sitting next to each other. Tetra's head was right next to Lisa's, and both were huddled behind Lisa's flame-red laptop. Despite Lisa's characteristic expressionlessness, there was an air of excitement. Tetra looked up and noticed me.

"Come here, come here, come here!" she half shouted, waving me over. She rushed to my side, enveloping me in a cloud of sweet scent. She grabbed my arm and pulled me over to the table.

"This is so cool, check it out!" she said, pointing at the computer screen. "Lisa got the pseudocode to actually run!"

"Actually run? What do you mean?" I asked.

I looked at the screen, which showed computer code for the LINEAR-SEARCH procedure we had talked about the other day.

"Why is part of it flashing?" I asked.

"It's showing what the computer's doing! See?"

Tetra started to point to the screen, but Lisa grabbed her arm with an uncanny swiftness.

"No," Lisa said.

"Oops," Tetra said. "No touching the screen. Sorry, I forgot."

In her deference to Lisa, for a moment Tetra seemed like the underclassman here.

"I think I see what's going on," I said. "It's marking the line in the pseudocode that's currently executing. The highlighted line changes as the program runs to show what's happening."

"That's right! And look, this table shows the variables! So, like, you can see what the value of k is at any time."

Tetra pointed at the screen, being careful not to touch it.

I watched as the $k \leftarrow k+1$ line executed and the k in the variable display increased from 379 to 380. It was fascinating.

"But why's this computer so slow?" I asked.

"It's not!" Lisa said, seemingly offended.

"We're slowing it down on purpose," Tetra said. "Lisa, show him what it looks like at full speed."

Lisa muttered something as she tapped some keys. When she stopped, the highlighted line started moving so fast I couldn't follow it. By the time I looked back at the values table, the previous k of 380 had become 22,000, then 23,000, then 24,000... The hundreds digits and below were moving so fast they couldn't be read.

"This is LINEAR-SEARCH, right? What value for n did you use?"

"Around one million, I think."

"One *million*?"

Lisa coughed softly. "1,048,576," she said. "The twentieth power of two."

"And you're searching from among that many numbers?"

"All the values in sequence A are 1," Tetra said, "but we're searching for a 0. So the procedure will end without finding the target value. Seems kind of mean to make the computer search for something we know isn't there, but it makes for a good experiment, so..." Tetra shrugged her shoulders.

I turned back to the screen.

"What did you mean when you said Lisa got the pseudocode to run? How, exactly?"

"I don't really understand the details, but she was able to input the pseudocode into the computer line by line, and use a program that turns the algorithm into a program that could do what the pseudocode said to do. A program that the computer can run. Did I get that right, Lisa?"

Lisa nodded.

"A program that creates programs?" I asked.

"An interpreter," Lisa said in that husky voice of hers, then turned back to another burst of silent machine-gun typing.

I looked at Lisa, even more impressed than before. She didn't just know how to use computers, she knew how to program them.

"Wow," I said.

"You want some serious wow?" Tetra half whispered. "Check out her keyboard."

I bent over for a closer look. The keys on Lisa's keyboard were the same red as the computer's case. Solid red. As in—

"Your keys are... blank? No letters, no numbers, no nothing?"

"No need," she said. "I never look at them."

4.3.2 The Dice Game

"Hey, let's do Yuri's dice game!" Tetra said.

She described it to Lisa, who rapidly typed up some pseudocode.

```
procedure DICE-GAME()
    a ← RANDOM(1, 6)
    b ← RANDOM(1, 6)
    if a > b then
        return ⟨Alice wins⟩
    else-if a < b then
        return ⟨Bob wins⟩
    end-if
    return ⟨Tie⟩
end-procedure
```

The dice game.

"What's this RANDOM(1, 6) line?" I asked.

"Dice," Lisa rasped.

"Ah, I get it. It's a function that returns a random number from 1 to 6, just like throwing a die, right?"

Lisa nodded, then ran the DICE-GAME program several times.

```
DICE-GAME() [↵]
⇒ ⟨Alice wins⟩

DICE-GAME() [↵]
⇒ ⟨Alice wins⟩

DICE-GAME() [↵]
⇒ ⟨Bob wins⟩

DICE-GAME() [↵]
⇒ ⟨Tie⟩

DICE-GAME() [↵]
⇒ ⟨Bob wins⟩
```

"Cool, right?" Tetra said. "We store Alice's roll in variable a and Bob's roll in variable b, then we compare the two to determine the winner. You get an idea, write it down in words in the form of a program, and the computer runs it. I love it!"

4.3.3 Roulette

"Hey, can you try something else?" I said.

I described Yuri's idea for a 36-sided roulette wheel, and asked Lisa if she could make a program that did the same thing. She only had one question.

"How do you spell 'roulette'?"

"R-O-U-L-E-T-T-E," Tetra whipped off.

It wasn't long before Lisa had a program ready to go.

```
procedure ROULETTE-GAME()
    r ← RANDOM(1, 36)
    if r ⩽ 15 then
        return ⟨ Alice wins ⟩
    else-if r ⩽ 30 then
        return ⟨ Bob wins ⟩
    end-if
    return ⟨ Tie ⟩
end-procedure
```

Roulette game equivalent to the dice game.

"This time we only need to call RANDOM once," Tetra said.

"But with a different range," I said. "RANDOM(1, 36) is like a single spin of a roulette wheel that returns a number from 1 to 36, and we store that integer value in a variable r. If that value is 15 or less then Alice wins, or Bob wins if it's 30 or less instead, and the game is a tie otherwise."

We now had two programs, DICE-GAME and ROULETTE-GAME. They worked in different ways, but both had the same results: a $\frac{5}{12}$ probability of an Alice win, a $\frac{5}{12}$ probability of a Bob win, and a $\frac{1}{6}$ probability of a tie.

"You sure can make these programs fast," Tetra said.

I nodded. "Yeah, amazing, isn't she?"

"Gah!" Lisa yelled.

We turned to see our resident math genius—seemingly having appeared from nowhere—standing behind Lisa and tousling her hair.

"Simulations?" Miruka said, peering over Lisa's shoulder.

"Enough, Miruka," Lisa mumbled.

4.4 Destruction of Certainty

4.4.1 Defining Probability

After playing with Lisa's programs for a while, Miruka, Tetra, and I started talking about math. Lisa remained near us, typing silently, but I couldn't be sure if she was listening or not.

Tetra opened with her question from lunchtime.

"—So anyway, I'm still not quite comfortable with 'same certainty.'"

"Hmph," Miruka grunted, closing her eyes.

A cool post-rain breeze blew in from the open window, along with faint cries from some sports team practicing on a far-off athletic field.

I observed Miruka in her silence—her metal-frame glasses, the ripples in her gorgeous black hair, the lines of her face, the elegance of her poise... But Miruka's attractiveness went far beyond her looks. Her knowledge was both broad and deep. She demonstrated freedom in her thought, and audacity in her decision-making. She used her abilities skillfully, and soared on wings of thought. It was no wonder she was an idol to Yuri, and to Tetra, and even to my mother.

Miruka opened her eyes.

"Let's define probability," she said.

"Sure!" Tetra replied.

"You, Tetra. Give it a go."

"Huh? You want *me* to give the definition?"

"I do. If you have doubts about something, trying to define it is a good first step."

"Um, okay. Well... Probability is..." Tetra paused to gather her thoughts, then began again. "Probability is the number of cases of some event you're interested in divided by the total number of possible cases."

"Not bad," Miruka said, then turned to me. "You. Counterpoint."

"I think that pretty much sums it up," I said. "I think you need to say that each event occurs with the same certainty, though. If you want the ratio of events of interest to all possible events to give a true probability, then you need to know that all events are equally likely."

"Good point!" Tetra said. "I agree, we need that condition."

"Except that you're still unsure about what 'same certainty' is in the first place, right?" I said.

"I guess I'm still not quite there."

"So are you really ready to make that part of your definition?"

"No, I suppose not. So maybe I should make a new definition, one that doesn't mention same certainties! But, is such a thing even possible?"

Tetra looked at me. I looked at Miruka.

"Well? *Is* such a thing possible?"

"It is," Miruka said. "So far as Tetra's question is concerned, yes, we can mathematically define probability without using the words 'equal certainty.'"

"So what do we use in its place?" Tetra asked.

"Axioms," Miruka said. "We'll provide axioms that allow the definition, and call whatever fulfills those axioms 'probability.' That's how probability is defined in mathematics. For convenience, how about we call probability defined using axioms 'axiomatic probability.'"

"That sounds different from what I learned in school, that probability is defined as a ratio of cases. Is that somehow wrong, mathematically speaking?"

"It's not wrong. In fact, it's what the mathematician Laplace used, a culmination of classical probability theory. There are some problems that classical probability can't handle, but it doesn't contradict axiomatic probability in any way."

Tetra looked relieved. "So the probability that I know—classical probability, I guess that is—still works."

"Mostly. But when we talk about defining probability in modern mathematics, we should use the axiomatic approach. Classical probability is just a special case of axiomatic probability."

Tetra brightened. "And if I learn the axioms of probability, I'll understand why the faces on dice come up with equal certainty?"

"Not quite," Miruka said. "No matter how well you know the probability axioms, they won't tell you that." Tetra deflated, but Miruka carried on. "If you want to precisely know the probability of dice throws, you need to throw dice. Throw them many times, and record occurrence frequencies. Doing that is called 'statistical probability.'"

"*Another* kind of probability?"

"Okay, let's summarize."

4.4.2 Kinds of Probability

"So, the three main kinds of probability," Miruka said.

She raised a finger.

"The first, as I said, we'll call axiomatic probability. This is probability as defined using the probability axioms. Its characteristics are determined using axioms, and whatever fulfills the axioms is defined to be probability. This is the definition of probability that's used in modern mathematics."

Miruka raised a second finger.

"Classical probability determines probabilities using ratios of cases. You start with a set of cases with equal certainty of occurrence, and represent probabilities as the number of cases of interest divided by the total number of cases. In other words, as a ratio of cases. This is the kind of probability you study up through high school. It's compatible with axiomatic probability, and has the advantage of being intuitive and easy to understand, but it has limited applications."

She raised a third finger.

"Statistical probabilities are determined using ratios of occurrence frequencies. This kind of probability is based on actual counts of how many times an event has actually occurred. You examine a set of occurrences and count how many times an event of interest happened, then take the ratio of those occurrences versus all the events you saw. You use past events to predict future ones, which is a good approach to take when it's hard to theoretically consider event causes. For example, how likely you are to be in a traffic accident within the next year."

4.4.3 Mathematical Applications

"I didn't realize there were so many kinds of probability," Tetra said.

"There's only one kind in modern mathematics—axiomatic probability," Miruka said. "The classical and statistical approaches are alternative approaches."

"Classical probability has the concept of same certainty?" Tetra asked.

"It does, in that you determine beforehand events with same certainty."

"And that's okay?"

"Sure. You need premises to some extent to talk about anything."

"But what if, like, the dice are loaded? Then you can't say that their faces will come up with the same certainty."

"Of course not."

"So doesn't that mean that we need some kind of precise mathematical definition of 'same certainty'?"

"Hmph. Let's think about this a little deeper." Miruka held up an imaginary die between her thumb and forefinger. "Say this is a loaded die. Where's the mistake if you use it in a probability problem?"

"There's a mistake?" Tetra asked.

"There is, but it isn't in the math. The mistake would be to use that die as an example of 'same certainty.' In other words, the problem isn't in the math, it's in your *application* of the math."

Tetra's eyebrows pulled together. "Hmmm..."

"Mathematics isn't concerned with whether each die face actually comes up with the same certainty. Mathematics just asks, *if the faces do come up with the same certainty*—if you're throwing what we call a 'fair die'—then what happens?"

Tetra bit a nail. She didn't look satisfied with Miruka's response.

"Doesn't that sort of feel like cheating?" she said. "Don't we want to know what will really happen? Isn't that what we want math to tell us?"

Miruka touched a finger to the rim of her glasses. "It's more like, given some conditions, math tells us what they imply."

"Well yeah, but—"

"It's the same thing. Mathematics is the study of what we can assert after specifying any preconditions. Anything beyond those boundaries is your responsibility."

"Huh," Tetra said.

"Tell you what," Miruka said. "Let's consider your doubts from each of the three perspectives."

4.4.4 Three Perspectives

"So here's your question," Miruka said, writing on a stray piece of paper:

What does it mean to say that the faces on a die will come up with equal certainty?

"From the perspective of axiomatic probability," she said, " 'same certainty' means that the probabilities assigned to each face are equal. In other words, we use the notion of probability as defined by axioms to say what 'same certainty' means. If we're considering a die to which we've assigned the same probability to each face, then we can say that each face will come up with the same certainty. We can say that we've modeled a fair die. Good so far?"

Tetra nodded, and Miruka continued. "Okay, on to the perspective of classical probability. This viewpoint is based on the condition that the faces of a fair die come up with the same certainty. It does not provide any description of what same certainty is, but it considers probability to be a ratio of cases when all those cases are of same certainty. In other words, we're saying that each face comes up with the same probability *if* each face is rolled with the same certainty. Still following?"

Tetra nodded again.

"Okay, the statistical probability perspective, then. In this case, you roll the die over and over again. If you've done that, and each face came up with approximately the same frequency, then you can say that each face has the same certainty. Of course, in this case you have other questions to answer, like how many times you need to throw the die, and how to handle frequency variations."

4.5 AXIOMATIC DEFINITION OF CERTAINTY

4.5.1 Andrey Kolmogorov

"I guess axiomatic probability is the one I understand the least," Tetra said. "Can you give me the details?"

"We start with some propositions called the probability axioms," Miruka said. "Whatever fulfills these probability axioms is what we'll call probability. We can also call it axiomatic probability, or probability from axiomatic principles."

"This sounds vaguely familiar..."

"I hope so. We did the same kind of thing when we defined groups.[1] We also used ring axioms to define rings, and field axioms to define fields. We used the Peano axioms to define the natural numbers, and we defined formal systems using the axioms of formal systems.[2] And now, we're using the axioms of probability to define probability." Miruka stood from her seat, raised a finger, and continued. "The Russian mathematician Andrey Nikolaevich Kolmogorov proposed the first axiomatic definition of probability in 1933."

"Yeah?" I said. "We didn't have an axiomatic definition until the twentieth century?"

Miruka gave a brief nod. "Kolmogorov was both a great mathematician, and a great teacher."

4.5.2 Sample Spaces and Probability Distributions

"So, axiomatic probability," Miruka said. "To prepare, we need to talk about sample spaces and probability distributions. A sample space is a set of basic events. A probability distribution is a function from a subset of the sample space to the real numbers. If these don't meet the probability axioms, then—"

"Wait, wait!" Tetra said, holding up both hands as if to restrain Miruka. "I'm not able to visualize any of this. What are you talking about?"

"Then we'll start with an example," Miruka said, pulling one of my notebooks toward her. "Say you throw a die. Let's consider a set omega, like this. And no, Tetra, it doesn't matter what we name it."

$$\Omega = \{\boxed{\cdot}, \boxed{\cdot\cdot}, \boxed{\cdot\cdot\cdot}, \boxed{::}, \boxed{:\cdot:}, \boxed{:::}\}$$

"This equation says Ω is a set with six elements, which we'll consider to be all possible dice rolls from 1 to 6. This is our sample space. Remember, the sample space Ω covers all possible dice throws. Conversely, there are no dice throws that are not elements of Ω—you can't roll a 0, and you can't roll a 7. In other words, the sample space has no leaks. Also, the elements in Ω cannot occur simultaneously—you can't simultaneously throw both a $\boxed{\cdot}$ and a $\boxed{\cdot\cdot}$ using one die. In other words, there are no dupes."

---

[1] See *Math Girls 2: Fermat's Last Theorem*.
[2] See *Math Girls 3: Gödel's Incompleteness Theorems*.

Tetra paused to scan our faces and make sure we were keeping up. We were.

"Next up is the Pr function. Pr is a function from a subset of Ω to the real numbers. For example, we can define a correspondence like this."

s	{⚀}	{⚁}	{⚂}	{⚃}	{⚄}	{⚅}
Pr(s)	$\frac{1}{6}$	$\frac{1}{6}$	$\frac{1}{6}$	$\frac{1}{6}$	$\frac{1}{6}$	$\frac{1}{6}$

"This table shows that the function Pr returns a value of $\frac{1}{6}$ when passed the set { some die face }. We're going to say it's this function that determines probabilities when we throw dice, and we're going to call this function the probability distribution for throwing dice. Also, we're going to call the real value $\Pr(\{x\})$ the probability that the x face comes up. We can write that as equations instead of using a table."

$$\Pr(\{⚀\}) = \tfrac{1}{6} \quad \Pr(\{⚁\}) = \tfrac{1}{6} \quad \Pr(\{⚂\}) = \tfrac{1}{6}$$
$$\Pr(\{⚃\}) = \tfrac{1}{6} \quad \Pr(\{⚄\}) = \tfrac{1}{6} \quad \Pr(\{⚅\}) = \tfrac{1}{6}$$

"We have to define the probability distribution Pr such that each of {⚀} through {⚅} return a value 0 or greater. In other words, no element in the sample space can give a negative or an undefined value. In the case we're talking about here all the probabilities are the same, but that's not a requirement. What *is* required is that the sum of all probabilities from the probability distribution Pr add up to 1. We'll talk more about the conditions the probability distribution fulfills later."

$$\Pr(\{⚀\}) + \Pr(\{⚁\}) + \Pr(\{⚂\}) + \Pr(\{⚃\}) + \Pr(\{⚄\}) + \Pr(\{⚅\}) = 1$$

"Well, what do you think, Tetra? Can you become friends with sample space Ω and probability distribution Pr?"

"Um, sure. I think I pretty much get it. Sample space Ω is the set of all possible events, and probability distribution Pr is a function for finding probabilities."

"And we use the set Ω and the function Pr to represent probability," I added.

"Good enough," Miruka said. "Okay then, a quiz to test our knowledge. What does this represent?"

$$\Pr(\{\boxdot\})$$

"That's, um, a dice throw that resulted in a ⚃," Tetra said.

"Wrong."

"No?"

"You're missing something. This is the *probability* of a dice throw resulting in a ⚃."

Statement	←----→	Meaning
$\{\boxdot\}$	←----→	A roll of ⚃
$\Pr(\{\boxdot\})$	←----→	Probability of rolling ⚃

"Right, right. Got it."

Listening to Miruka and Tetra's back-and-forth, I started thinking about dialogue and the important role it plays in learning. When a teacher has multiple students, those students will understand what's being taught to different extents. So is it really possible for honest questions and answers to arise in a school classroom? Is it really possible for dialogue to lead to deep understanding? *Making that possible is precisely the job of a teacher, I guess.*

"Oh, one question!" Tetra said, with a hand on her head. "I'm not sure what the difference is between a probability and a probability distribution."

Miruka replied, "$\Pr(\{\boxdot\})$ is the probability of rolling a ⚃. In this example, it's the real value $\frac{1}{6}$."

"I get that," Tetra said.

"Pr is the probability distribution, a function that shows the relation between what can happen and the probability that those things happen. So one is a real number, the other is a function."

"I understand how they're different in that sense, but I'm not sure where the 'distribution' part of 'probability distribution' comes from."

"Hmph," Miruka said, and paused to think. After a few finger taps, she said, "A minute ago I told you that all the probabilities have to add up to 1. Well, let's look at it from the other direction. Namely, how the probability distribution Pr *distributes* the probability 1 over the sample space Ω."

"It distributes... Oh! I get it! This function called Pr chops the probability of 1 up into little pieces, and, like," Tetra mimed what looked like a flower girl tossing petals at a wedding, "sprinkles them out among each of the events."

"Close enough," Miruka said. "The probability distribution function does exactly that—it distributes probabilities. It says where there's a high probability, and where there's a low one. Where the hills are, and where the valleys are. The probability distribution sets all the scenery."

"And a fair die creates plains," I said.

4.5.3 The Probability Axioms

"Let's talk about Kolmogorov's probability axioms."

Probability axioms

Let Ω be a set with subsets A and B, and let Pr be a function from the set of subsets of Ω to the real numbers. Further, assume that Pr fulfills the following axioms P1, P2, and P3:

Axiom P1 $0 \leqslant \Pr(A) \leqslant 1$

Axiom P2 $\Pr(\Omega) = 1$

Axiom P3 If $A \cap B = \{\}$, then $\Pr(A \cup B) = \Pr(A) + \Pr(B)$

Then,

- we call Ω the *sample space*,
- we call subsets of Ω *events*,
- we call Pr the *probability distribution*, and
- we call the real value $\Pr(A)$ the *probability* of the occurence of A.

"That..." Tetra stammered, "that doesn't look *at all* like probability."

Miruka narrowed her eyes. "Well what does it look like, then?"

"Sets, maybe?"

"As it should. Axiomatic probability uses sets and logic. After all, it would be circular logic to define probability using probability."

"I guess it would, huh?"

4.5.4 Subsets and Events

Tetra reviewed the probability axioms.

"Okay," she said, "one of the probability axioms says we call subsets of Ω 'events,' but what's an event here?"

"Something that will happen," Miruka said. "Or something that happened, or something that can happen."

"And the probability distribution Pr is a function for obtaining probabilities of those events?" I said.

Miruka nodded. "Tetra, do you understand the definition 'A is a subset of Ω'?" she said.

"It means that A is a part of Ω... I think," Tetra mumbled.

This time she shook her head. "Not good enough. You've got the gist of things, but that doesn't make for a definition. Better to say, 'all elements of A are also elements of Ω.' You—" she said, pointing at me, "give me a subset of this Ω."

$$\Omega = \{\boxed{\cdot}, \boxed{\cdot\cdot}, \boxed{\cdot\cdot\cdot}, \boxed{::}, \boxed{:\cdot:}, \boxed{:::}\}$$

Easy enough.

"How about something like this?"

$$A = \{\boxed{\cdot\cdot}, \boxed{:\cdot:}, \boxed{:::}\}$$

"Take any element in this set A, $\boxed{\cdot\cdot}$ or $\boxed{:\cdot:}$ or $\boxed{:::}$, and it will also be an element of Ω. So, by definition, A is a subset of Ω. We write it like this."

$$A \subset \Omega$$

"Conceptually, you can think of it as meaning 'A is contained in Ω.'"

"Draw a diagram," Miruka said.

"Yeah, sure. It looks something like this."

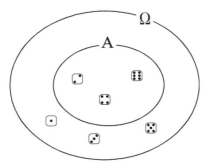

Set A is a subset of set Ω
Set A is included in set Ω
$A \subset \Omega$.

"Okay, I understand the definition," Tetra said. "But what kind of event does this set A represent?"

"What do you think it represents?," I asked.

"Is it... rolling an even number?"

"It is!" I said. "There are plenty of other die-throw events we can think of. In fact, there's 2^6 of them in all."

"And don't forget," Miruka said, "Ω is a subset of itself, as is the null set { }." She paused to stretch before continuing. "Okay, enough warmup. Let's go back to Kolmogorov's probability axioms, and take a more careful look. It's those axioms that define mathematical probability."

4.5.5 Probability Axiom P1

Probability Axiom P1

Probability Axiom P1 $0 \leqslant \Pr(A) \leqslant 1$

"Somebody tell me what this inequality means," Miruka said.

$$0 \leqslant \Pr(A) \leqslant 1$$

"That the probability of event A is 0 or greater and 1 or less?" Tetra said. "But how do we know this Pr function fits in the inequality?"

"The question doesn't make sense. We aren't trying to say that Pr fulfills the inequality. We're saying that in order for Pr to be called a probability distribution, it *must* fulfill the inequality."

"Oh."

"We're using mathematical statements to establish some conditions that this function Pr has to meet. In other words, we're setting some restrictions on Pr, and saying that only functions behaving within those restrictions can be called probability distributions. Standard practice for an axiomatic definition."

Tetra started writing in her notebook.

"This restriction is the requirement for a probability distribution..."

"Careful now," Miruka said. "Axiom P1 is the *first* requirement."

"Oops!" Tetra said, and made some corrections.

"By the way, remember how Ω is a subset of itself?" Miruka's voice lowered to nearly a whisper. "So what kind of concept does Ω represent?"

"Um, the set of everything that can happen?" Tetra said. "No, wait! Is it what will definitely happen?"

"Right, Ω is called the 'certain event,' and this is what axiom P2 defines."

Probability Axiom P2

Probability Axiom P2

Probability Axiom P2 $\Pr(\Omega) = 1$

"This axiom P2 is another restriction on probability distributions," Miruka said. "So again, a function that doesn't meet this restriction isn't a probability distribution."

Tetra's eyes widened.

"And it says that the probability of *something* happening is 1, right?" she said.

"That's not wrong, but let's not clump things together so much. All axiom P2 says is that the probability of some event occurring is 1, which we can interpret as saying that some event occurs with certainty."

"In the sense of mathematical applications," I added.

"Or, you can think of it as the mathematical concept of a 'certain event' representing the real-world concept of 'definitely happens.'"

$$\Pr(\{\boxed{\cdot},\boxed{\cdot\cdot},\boxed{\cdot\cdot\cdot},\boxed{::},\boxed{:\cdot:},\boxed{:::}\}) = 1$$

"Pop quiz," Miruka said. "Can we find a value for the probability $\Pr(\{\})$?"

"I know this!" Tetra said. "It's 0, right?"

"By which axiom?"

"Um, because the probability of definitely not happening is 0, isn't—"

"Bzzzt. We're trying to axiomatically define probability here. You *have* to use axioms."

"Yeah, but from axiom P2 we know that $\Pr(\{\boxed{\cdot},\boxed{\cdot\cdot},\boxed{\cdot\cdot\cdot},\boxed{::},\boxed{:\cdot:},\boxed{:::}\}) = 1$, and axiom P1 says that all probabilities have to be from 0 to 1, so doesn't that mean we can say $\Pr(\{\}) = 0$?"

"Nope. Axioms P1 and P2 alone aren't enough."

"So what's missing?"

"Axiom 3."

Probability Axiom P3

Probability Axiom P3

Axiom P3 If $A \cap B = \{\}$, $\Pr(A \cup B) = \Pr(A) + \Pr(B)$

"Let's do a quick review of set operations," Miruka said. She turned to Tetra. "What does $A \cap B$ give you?"

"A set created from elements that are in both A and B," Tetra said.

"That's right. We call this the intersection of the sets. Here's an example."

$$\{\boxed{\cdot},\boxed{\cdot\cdot},\boxed{\cdot\cdot\cdot},\boxed{::}\} \cap \{\boxed{\cdot\cdot},\boxed{::},\boxed{:::}\} = \{\boxed{\cdot\cdot},\boxed{::}\}$$

"Here's another example, one where the intersection is the empty set."

$$\{\boxed{\cdot},\boxed{\cdot\cdot},\boxed{\cdot\cdot\cdot}\} \cap \{\boxed{\cdot\cdot},\boxed{\cdot\cdot\cdot},\boxed{:::}\} = \{\,\}$$

"Sets whose intersection is the empty set are called disjoint sets. If you consider them as events, two disjoint sets are called mutually exclusive events."

Tetra muttered as she jotted down these new terms.

"Disjoint sets... Mutually exclusive..."

"Moving on," Miruka said. "What's $A \cup B$?"

"A set of elements belonging to either A or B!"

"Either–or?" Miruka said, raising an eyebrow.

"Er, no! I should have said, to *at least one of* A or B. It will still be in the resulting set if it's in both."

"Better. $A \cup B$ is called the union of A and B. Here's an example."

$$\{\boxed{\cdot},\boxed{\cdot\cdot},\boxed{\cdot\cdot\cdot}\} \cup \{\boxed{\cdot},\boxed{\cdot\cdot},\boxed{\cdot\cdot},\boxed{\cdot\cdot\cdot}\} = \{\boxed{\cdot},\boxed{\cdot\cdot},\boxed{\cdot\cdot},\boxed{\cdot\cdot\cdot},\boxed{\cdot\cdot\cdot}\}$$

"Okay, now we're ready," Miruka said. "Do you understand axiom P3 now?"

Axiom P3 If $A \cap B = \{\,\}$, $\Pr(A \cup B) = \Pr(A) + \Pr(B)$

"Um, maybe?" Tetra said.

"So what should you do?" Miruka shot back.

"I, uh..."

"Examples are...?"

"Oh! 'Examples are the key to understanding'! Okay, an example of sets whose intersection is the empty set would be something like this."

$$A = \{\boxed{\cdot},\boxed{\cdot\cdot},\boxed{\cdot\cdot\cdot}\}$$
$$B = \{\boxed{\cdot\cdot},\boxed{\cdot\cdot\cdot},\boxed{:::}\}$$

"Then axiom P3 says that this should be true, right?"

$$\Pr(\{\boxed{\cdot},\boxed{\cdot\cdot},\boxed{\cdot\cdot},\boxed{\cdot\cdot\cdot},\boxed{\cdot\cdot\cdot},\boxed{:::}\}) = \Pr(\{\boxed{\cdot},\boxed{\cdot\cdot},\boxed{\cdot\cdot\cdot}\}) + \Pr(\{\boxed{\cdot\cdot},\boxed{\cdot\cdot\cdot},\boxed{:::}\})$$

4.5.6 Not Quite There

"Good. Get it now?" Miruka said.

Tetra slouched a bit.

"Something about axiom P3 isn't quite clicking for me yet. I mean, I understand the equations, but I'm not quite sure why it's important."

Miruka laughed. "I love the things you get hung up on, Tetra."

"Yeah?"

"Let's change our perspective. Axiom P3 gives us a guideline for finding the probability of an event, namely that we should divide things up into mutually exclusive events."

Tetra tilted her head.

"Look at axiom P3 again. It says that for mutually exclusive events A and B, we have this."

$$\Pr(A \cup B) = \Pr(A) + \Pr(B)$$

"In other words, if $A \cap B = \{\}$ for events A and B, then we can get the probability $\Pr(A \cup B)$ from the sum of $\Pr(A)$ and $\Pr(B)$. Simply put, the probability of the union is the sum of the probabilities."

"Oh..."

"The probability axioms give us a good taste of what probability is all about," she said. "Probability is *normalized quantities*."

Miruka stood from her seat.

"Say you want to find the probability of some event," she continued. "All you have to do is chop it up into mutually exclusive events, and find the probabilities of each of those."

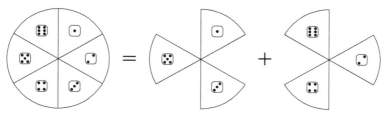

Separation and combination of mutually exclusive events.

"Probability is all about breaking up and combining—there's no increasing or decreasing. It's a *quantity*. Also, the quantity of all the parts together will always be 1. That's *normalization*. That's why I say that probability is about normalized quantities. Quantities where the probability distribution is divvying up a 1 among the elements in the sample space."

4.5.7 Even Odds

Miruka returned to her seat beside Tetra.
"A problem for you," she said.

Problem 4-1 (probability of rolling an even number)

Let Ω be a sample space, where

$$\Omega = \{\boxed{\cdot}, \boxed{\cdot\cdot}, \boxed{\cdot\cdot\cdot}, \boxed{::}, \boxed{:\cdot:}, \boxed{:::}\}.$$

Let Pr be a probability distribution, where

$$\Pr(\{\,d\,\}) = \frac{1}{6} \qquad (d = \boxed{\cdot}, \boxed{\cdot\cdot}, \boxed{\cdot\cdot\cdot}, \boxed{::}, \boxed{:\cdot:}, \boxed{:::}).$$

Given the above, find the value of

$$\Pr(\{\boxed{\cdot\cdot}, \boxed{::}, \boxed{:::}\}).$$

"In other words, you want the probability of rolling an even number," Tetra said.

"Yep."

"Then the value should be $\frac{1}{2}$."

"Yep."

"So, yeah..."

"What I want is for you to show me where that comes from. You need to derive it using the probability axioms. That's what's needed if we're going to base all this on an axiomatic definition."

"Oh, but... I..."

"Next batter," Miruka said, looking at me. "What do you think?"

"Let's see..." I went back over the axioms, trying to see how they forged a path to the answer. "I think we can just apply axiom P3.

Yeah, that's it. We break things up into mutually exclusive events, like this."

$$\Pr(\{⚀,⚂,⚅\})$$
$$= \Pr(\{⚀\}) + \Pr(\{⚂,⚅\}) \quad \text{(Axiom P3) divide into mutually exclusive events } \{⚀\} \text{ and } \{⚂,⚅\}$$
$$= \frac{1}{6} + \Pr(\{⚂,⚅\}) \quad \text{(Probability distribution) use } \Pr(\{⚀\}) = \frac{1}{6}$$
$$= \frac{1}{6} + \Pr(\{⚂\}) + \Pr(\{⚅\}) \quad \text{(Axiom P3) divide into mutually exclusive events } \{⚂\} \text{ and } \{⚅\}$$
$$= \frac{1}{6} + \frac{1}{6} + \Pr(\{⚅\}) \quad \text{(Probability distribution) use } \Pr(\{⚂\}) = \frac{1}{6}$$
$$= \frac{1}{6} + \frac{1}{6} + \frac{1}{6} \quad \text{(Probability distribution) use } \Pr(\{⚅\}) = \frac{1}{6}$$
$$= \frac{3}{6} \quad \text{add}$$
$$= \frac{1}{2} \quad \text{simplify}$$

I reviewed what I had written, and nodded.
"Yeah, this should give us a probability of $\frac{1}{2}$."
"Well done," Miruka said.

Answer 4-1 (probability of rolling an even number)

$$\Pr(\{⚀,⚂,⚅\}) = \frac{1}{2}$$

"We can find that $\Pr(\{\}) = 0$ the same way," I said.

$$\Pr(\{\boxdot,\boxdot,\boxdot,\boxdot,\boxdot,\boxdot\}) = 1$$ (Axiom P2) probability of a certain event is 1

$$\Pr(\{\,\} \cup \{\boxdot,\boxdot,\boxdot,\boxdot,\boxdot,\boxdot\}) = 1$$ union of the empty set and the whole set is the whole set

$$\Pr(\{\,\}) + \Pr(\{\boxdot,\boxdot,\boxdot,\boxdot,\boxdot,\boxdot\}) = 1$$ (Axiom P3) divide into mutually exclusive sets $\{\,\}$ and $\{\boxdot,\boxdot,\boxdot,\boxdot,\boxdot,\boxdot\}$

$$\Pr(\{\,\}) + 1 = 1$$ (Axiom P2) probability of a certain event is 1

$$\Pr(\{\,\}) = 0$$ subtract 1 from both sides

"So *that's* how you do it," Tetra said. "I think I'm getting used to the probability axioms, but now I have a new question—why would we want to use axioms to define probability in the first place?"

"They allow us to talk about probability just by creating an Ω and a Pr, in other words a sample space and a probability distribution."

"Which means...?"

"Which means we can make precise mathematical descriptions of things like loaded dice, or coins landing on their side."

My ears pricked up.

"What was that about coins?" I said.

4.5.8 Loaded Dice and Edgy Coins

"Let's start with a loaded die," Miruka said. "First, we need a sample space and a probability distribution that describe it. Something like this."

Sample space Ω

$$\Omega = \{\boxdot,\boxdot,\boxdot,\boxdot,\boxdot,\boxdot\}$$

Probability distribution Pr

s	$\{\boxdot\}$	$\{\boxdot\}$	$\{\boxdot\}$	$\{\boxdot\}$	$\{\boxdot\}$	$\{\boxdot\}$
Pr(s)	0.1651	0.1611	0.1645	0.171	0.1709	0.1674

"These numbers have to add up to 1, right?" Tetra said.

$$0.1651 + 0.1611 + 0.1645 + 0.171 + 0.1709 + 0.1674 = 1$$

"They do. For the edge-landing coin, it might be something like this."

Sample space Ω

$$\Omega = \{\,\text{Heads, Tails, Edge}\,\}$$

Probability distribution Pr

s	$\{\,\text{Heads}\,\}$	$\{\,\text{Tails}\,\}$	$\{\,\text{Tails}\,\}$
$\Pr(s)$	0.49	0.49	0.02

"A 0.02 probability of landing on its edge?" I said. "Sounds awfully high. I doubt I could flip a coin 100 times and get it to land on its edge twice."

"All that means," Miruka said, "is that this probability distribution isn't a good description of the real world. That doesn't say anything about its suitability as a probability distribution. All I want to do is use a probability distribution to mathematically represent a probabilistic phenomenon. If you want it to be real-world accurate, all you have to do is fiddle with the numbers."

"Fair enough."

4.5.9 Promises

"Modern probability theory starts with the probability axioms," Miruka said. "The real-world notions of what happens and the certainty with which they happen are represented by a set called the sample space and a function called the probability distribution."

"I'm happy to say that I'm finally getting to be on good terms with those words," Tetra said.

"Well, I hate to break up a happy friendship," Miruka said, "but we'll often be ignoring the sample space."

"Oh, really? Why?"

"Because a random variable and a probability distribution are enough to do what we want."

Tetra raised her hand.

"What's a random variable?" she asked.

"The library is closed!" came a call from the librarian's office—Ms. Mizutani's punctual call to leave.

"Time's up," Miruka said. "I'll tell you tomorrow."

Something tightened in my chest.

"Maybe you shouldn't make promises like that," I said.

"No promises, then. I'll tell you about it... eventually."

"Off to home, then!" Tetra said. "Let's go!"

"Actually, there's something I want to talk to you about," Miruka said.

Tetra froze. "Who...? Me?"

4.5.10 Coughs

Tetra scuttled off with Miruka, leaving me with Lisa. Walking to the station with the silent, expressionless girl was... well, awkward at best.

"You always carry your computer with you?" I asked in the cheeriest tone I could manage.

She nodded.

"You must be pretty attached to it."

"I like its keyboard," she said. She softly coughed and shifted her bag.

"What kind of keyboard is it?"

"Dvorak."

"What's that?"

"A Dvorak simplified keyboard." She coughed again.

I wasn't sure what she was talking about, so I let the subject drop.

"So you're Dr. Kurabikura's daughter?"

Another nod.

"Do you live near the Kurabikura Library?"

Nod.

I grimaced, realizing that I sounded like I was doing a background check.

"Does Miruka go to the Kurabikura Library much?" I asked.

"Miruka—" she started, then began coughing in earnest. Softly at first, then stronger, like something was stuck in her throat. It was painful to watch. She doubled over, both hands covering her mouth. I crouched down next to her.

"You okay?"

She nodded, eyes closed, but did not look okay at all. I hesitantly put a hand on her back. It was surprisingly cold. After a minute or two, the coughing subsided.

"Better now?"

Lisa stood back up.

"I'm not much of a talker," she said.

"Not that it's any of my business, but maybe you should cut back on cold drinks? You're chilled to the bone, and it looks like you've done a number on your throat."

I winced—I was giving her the same advice my mother would have. Even so, Lisa brightened a little.

"Maybe so," she said.

I even thought I saw a hint of a smile. Just a small one.

The theory of probability, as a mathematical discipline, can and should be developed from axioms in exactly the same way as geometry and algebra.

ANDREY KOLMOGOROV
Foundations of Probability Theory

CHAPTER **5**

Expected Values

> Fear of danger is ten thousand times more terrifying than danger itself.
>
> DANIEL DEFOE
Robinson Crusoe

5.1 RANDOM VARIABLES

5.1.1 *Mom*

"Whatcha studying?" my mother asked, barging into my room.

"Kinda busy here," I said. I knew her well enough to know she wasn't really interested in my studies. She wanted me to do something for her.

"Aww," she said, making her "disappointed" face. "I wanted you to help with dinner."

"Still busy."

"Seems like just yesterday you were always tugging at my apron, asking if you could help out."

"Seems like just yesterday when you last brought that up."

"This is why I should have had a girl." She wandered over to my bookshelf and scanned the titles. "Speaking of which, what happened to all those girlfriends of yours? Do you guys have anything planned? They haven't been over in so long. Did you chase them off?"

"We don't have any plans, and you wouldn't be involved even if we did."

"Is Miruka an only child?"

"She had a brother, but he died when she was in elementary school."

"Oh, my. Was he sick?"

"Don't know. Can you please leave now?"

Watching my mother retreat, I remembered Miruka's tears on the day she told us about her brother. How Tetra had wiped them away with a handkerchief when Miruka refused to wipe them away herself.

5.1.2 Tetra

The next day, Tetra was sitting alone in the library when I arrived after classes. She was sitting with her back to the door, possibly to enjoy the warm spring breeze blowing in from the window, and didn't notice me coming in. I walked up behind her (as quietly as possible, I'll admit), and said (perhaps just a little too close to her ear)...

"Hey, Tetra."

The resulting scream drew every eye in the place.

Yes, it was my own fault, and I knew exactly what would happen. I don't know what got into me.

"Don't sneak up on me like that!" Tetra said, feigning a slap with her notebook.

"Sorry about that," I said, sliding into the chair next to her. "Are you studying?"

"Yeah, doing some math. Remember how Miruka said something about random variables the other day? Well, I was going to wait for her to teach us about them, but then I remembered what you said."

"Remind me what that was."

"That I shouldn't just sit around, passively waiting for people to teach me things. That if I'm interested in something, I should pick up a book and study on my own."

"Well good for you."

"So I got a book, and gave it my best shot, but...well, learning something new on your own is hard. Like, here's what this book says about random variables."

She turned to a page in the book that was open in front of her, and pointed at a definition.

Definition of random variable

A *random variable* is a function from a sample space Ω to the real numbers $\mathbb{R}$.

"Interesting," I said.

"Maybe, but I keep staring and staring at this and it just doesn't seem to make sense. Especially this part about a random *variable* being a *function*. Isn't that confusing?"

"I guess that might seem a little strange, sure," I said, nodding. "But you're making a mistake if you just stopped there."

"I read a little further, but then I ran into this definition of expected values, and things just got worse."

Definition of expected value

The *expected value* $\mathsf{E}[X]$ of a random variable X is defined as

$$\mathsf{E}[X] = \sum_{k=0}^{\infty} c_k \cdot \Pr(X = c_k),$$

where

- $c_0, c_1, c_2, c_3, \ldots, c_k, \ldots$ are values that X can take, and
- $\Pr(X = c_k)$ is the probability that X equals c_k.

"This one is totally beyond me," Tetra said. "I gave up before I even figured out what this $\mathsf{E}[X] = \cdots$ means."

She splayed both arms on the table and collapsed in a pose of total defeat.

"I think it might be because you're trying to take the whole thing in at once," I said. "Just take your time and plug away at it. The math isn't going to run off anywhere. Just, you know, make friends with the new words one at a time."

Tetra, still crashed on the table, just sighed.

"Also, this particular book might be a tad above your level," I said. "But I'll bet it can't beat the two of us combined. Want to give it a shot?"

"You bet!" she said, perking up in her seat. "Um, if you have the time, of course."

5.1.3 Examples of Random Variables

"So first we want to become friends with this thing called a random variable. We want to be able to deal with problems involving probability. Things like rolling dice, or flipping coins, or drawing cards, right?"

"Sure," Tetra said, giving a large nod.

I opened her notebook to a fresh page.

"Here are some examples," I said, creating a list.

- Rolling a fair die.

- Flipping a weighted coin, which has probability 0.49 of coming up heads, and probability 0.51 of coming up tails.

- Trying to draw the 1 winning ticket out of a stack of 100.

"Sure, no problem," Tetra said.

"We want to *quantitatively* consider problems like this. A random variable is one of our best weapons for doing so."

"It's...a weapon?"

"Or a tool, if you prefer. We use variables all the time when solving math problems, right? Like when we say we'll use x to represent something. Then we can stick that x into an equation, and solve for it."

"Sure! We use x and y to represent things in word problems."

"We do. And random variables do something very similar. We use them to stand for other things in equations."

"Can you give me an examples?"

"I'll give you several. Like this."

- We can let a random variable X stand for the number rolled on a die.

- We can let a random variable Y stand for the number of heads in ten coin flips.

- We can let a random variable Z stand for the number of tickets drawn before drawing the winning ticket.

"That's it?" Tetra said. "That's all there is to them?"

"Pretty much. A random variable can be any real number that's determined through some number of trials. A dice roll is probably the most basic example."

"What's a trial?"

"Performing some action, like rolling a die."

Tetra mimed a dice throw as she thought on this.

"Huh. Any other examples with dice? I can't quite come up with one on my own."

"Sure. You could create..."

- A random variable that takes 100 times the value of a dice roll.

- A random variable that's 0 on an even dice throw, 1 on an odd throw.

- A random variable whose value increases by 100 on a throw of 4 or higher, and decreases by 100 on a throw of 3 or less.

"Oh, neat! So you can do calculations, and treat them differently in different cases, stuff like that."

"Right. All you have to do is state what value the variable will take when you declare it."

Tetra took a moment to consider my words, practically groaning with the effort. Long mathematical descriptions could sometimes confuse her, and she hadn't completely conquered her tendency to forget conditions when working problems, but she never failed to tackle new math with focused effort.

"Hey, I've noticed something!" she said, raising her hand. "I think I have a good idea of what random variables are all about, but why is the book using a capital X to represent them? Aren't variables normally lower case?"

"Huh, interesting point. Well, it's just a name, so really it can be anything—upper or lower case, Greek letters or Roman... But you're right, books on probability tend to write random variables using capital letters. I think it might be to distinguish between the specific values they can take, which are represented using lower-case letters."

"Hmm, okay. One more question! This definition for random variables says they're functions from a sample space Ω to the real numbers. I still don't see how a variable can also be a function."

There it is. Tetra's "I still don't get it." Her superpower. Her ability to tenaciously maintain that state.

"Okay, then let's take a look at how we determine the value of a random variable. For example, say you roll a ⚀. That's equivalent to specifying a point in the sample space, and it sets one random variable value to ⚀. So you can think about the random variable as associating one of its real-number values with the results of a die throw. We can mathematically represent that feature of random variables by saying they're a function from the sample space to the real numbers."

"Uh..."

"Too abstract? Okay, say you're playing a game where you roll dice for money."

The 100× game (random variable example)

Throw a die, and receive 100 times the value thrown in yen. We will use a random variable to track the prize money of $X(\omega)$ dollars, where ω is the value of the thrown die. The random variable $X(\omega)$ is a function from sample space $\Omega = \{⚀,⚁,⚂,⚃,⚄,⚅\}$ to the real numbers $\mathbb{R}$.

Throw (element ω in Ω)	⚀	⚁	⚂	⚃	⚄	⚅
Prize (value of $X(\omega)$)	100	200	300	400	500	600

"Okay, now this I think I can handle," Tetra said.

"Yeah, using a table to show the function mapping from the sample space to the real numbers should make it easier to conceptualize."

"It does!"

"You understand what the table means, right? We're setting ω to the number we throw, and using $X(\omega)$ to represent the reward. So if $\omega = \boxed{\cdot\cdot\cdot}$, then $X(\omega) = 300$. In other words, $X(\boxed{\cdot\cdot\cdot}) = 300$ yen. Nothing hard about that, right?"

"Nope! Throw the die, win some money. No problem."

"And that's all there is to it. So in this game, the random variable $X(\omega)$ acts like this."

- It's a *variable*, in the sense that it represents the prize money.

- It's a *function*, in the sense that it determines the prize money according to the die throw.

"Wow, I was way overthinking this. All we're doing is tying prize money to dice rolls, like this."

$$X(\boxed{\cdot}) = 100, \ X(\boxed{\cdot\cdot}) = 200, \ X(\boxed{\cdot\cdot\cdot}) = 300, \ \ldots, \ X(\boxed{::::}) = 600$$

"Yep, that's all."

"Sorry to ask another question, but sometimes you're writing the random variable X as $X(\omega)$. What's that all about?"

"We can write it that way to show that its value is determined by a sample space element ω, just like we often write functions as $f(x)$. If we roll a $\boxed{\cdot}$, it's convenient to use $X(\boxed{\cdot})$ to represent the value of the random variable. It's a different way of writing it, but it means the same thing as X."

5.1.4 Examples of Probability Distributions

Tetra turned back to her book and flipped through its pages.

"Okay, I think I'm pretty good with random variables now, but what about expected values? That's still beyond me."

"We can cover that too, but for now let's become better friends with probability distributions. Specifically, we should talk about the probability distributions of random variables. Simply put, they represent the probability that a random variable will take a specific value. Let's create an example from the 100× game we just looked at."

> **Example probability distribution for random variable X**
>
> | Value c that X can take | 100 | 200 | 300 | 400 | 500 | 600 |
> |---|---|---|---|---|---|---|
> | $\Pr(X = c)$ | $\frac{1}{6}$ | $\frac{1}{6}$ | $\frac{1}{6}$ | $\frac{1}{6}$ | $\frac{1}{6}$ | $\frac{1}{6}$ |

"What does this $\Pr(X = c)$ mean?" Tetra asked. "It seems kind of strange to have an equation in the parentheses there."

"The $X = c$ is just a restriction saying that X has to equal c. So $\Pr(X = c)$ means the probability that $X = c$."

"And what happened to the dice rolls?"

"We don't need to think about actual rolls when we're talking about probability distributions. Of course, there's a sample space floating around in the background somewhere[1], but we don't have to concern ourselves with it just now."

"We can do that?"

"In terms of the 100× game, it's like this. So long as we know the prizes and probabilities, we don't really need to think about what's rolled. The dice are beside the point, if we know the probability of winning a prize."

"Uh..."

"In other words, we're saying that instead of a *sample space* and a probability distribution, we can think in terms of *the value of a random variable* and a probability distribution."

Tetra nodded slowly as she digested this.

"Okay," she said, "so we just need to know what values the random variable can take, and at what probabilities. I'm good that far. But I don't see why we would want to look at things that way."

"You will eventually. For now, just remember what it is that makes up the probability distribution of a random variable—the values it can take, along with the probabilities that it takes those values."

---
[1] $\Pr(X = c)$ is taken to be $\Pr(\{\omega \in \Omega \mid X(\omega) = c\})$

5.1.5 So Many Words

"I've noticed something else," Tetra said while jotting down more notes. "I've been confusing three words."

"Which ones?"

"Probability, random variable, and probability distribution. Can you make sure I have them straight now? I need get these words down right if I'm going to figure all this stuff out."

Tetra showed me what she'd been writing:

- The *probability* of rolling ⚀ on a die is $\frac{1}{6}$.

- We can represent the winnings in the 100× game using a *random variable*.

- If we look at the *probability distribution* of a random variable X, we can know the probabilities at which X takes its values.

"Looks like you've got it now," I said. "By the way, good for you. When you don't understand something, you don't just say 'I don't understand,' you look for exactly what you don't understand, and why."

"Aww, cut it out," Tetra said, blushing and scratching her head.

5.1.6 Expected Values

"I think that pretty much covers random variables and probability distributions," I said.

Tetra nodded in satisfied agreement.

"I've become much better friends with both!"

"Okay, so now we're ready to move on to expected values."

"Starting with...just what *are* they?"

"Well, they're basically averages. The expected value of a random variable X is its average value."

"An average, as in an add-them-up-and-divide-by-the-count average? That's all the funky definition in this book is saying?"

"Yes, that kind of average, and yes, it's right there in the definition."

> **Definition of expected value**
>
> The *expected value* E [X] of a random variable X is defined as
>
> $$E[X] = \sum_{k=0}^{\infty} c_k \cdot \Pr(X = c_k),$$
>
> where
>
> - $c_0, c_1, c_2, c_3, \ldots, c_k, \ldots$ are values that X can take, and
> - $\Pr(X = c_k)$ is the probability with which X equals c_k.

"Um, I'm sure it is, but..."

"Don't worry, we can use our 100× game again as an example. We're letting the random variable X be the amount of the winnings. It's defined like this."

$$\text{Expected value of } X = E[X] = \sum_{k=0}^{\infty} c_k \cdot \Pr(X = c_k)$$

"Let's write out the sigma to better see what's going on. We'll keep the same values we used before: $c_0 = 100$, $c_1 = 200$, $c_2 = 300$, ..., $c_5 = 600$."

$$\begin{aligned}
E[X] &= \sum_{k=0}^{\infty} c_k \cdot \Pr(X = c_k) \\
&= \sum_{k=0}^{5} c_k \cdot \Pr(X = c_k) \quad \text{just need } c_0, c_1, c_2, \ldots, c_5 \\
&= 100 \cdot \Pr(X = 100) \\
&\quad + 200 \cdot \Pr(X = 200) \\
&\quad\quad + 300 \cdot \Pr(X = 300) \\
&\quad\quad\quad + 400 \cdot \Pr(X = 400) \\
&\quad\quad\quad\quad + 500 \cdot \Pr(X = 500) \\
&\quad\quad\quad\quad\quad + 600 \cdot \Pr(X = 600)
\end{aligned}$$

"Since the probability $\Pr(X = \text{winnings})$ will be $\frac{1}{6}$ in every case, this simplifies nicely."

$$= 100 \cdot \frac{1}{6}$$
$$+ 200 \cdot \frac{1}{6}$$
$$+ 300 \cdot \frac{1}{6}$$
$$+ 400 \cdot \frac{1}{6}$$
$$+ 500 \cdot \frac{1}{6}$$
$$+ 600 \cdot \frac{1}{6}$$

"All that's left are some simple calculations."

$$= \frac{100 + 200 + 300 + 400 + 500 + 600}{6}$$
$$= \frac{2100}{6}$$
$$= 350$$

"So the expected value for our winnings is 350 yen," I said.
"Hey, look at this!"

$$\frac{100 + 200 + 300 + 400 + 500 + 600}{6}$$

"We're adding up six numbers, then dividing by 6! That totally looks like an average! Not that we could see it in the definition."

"Sure, in the sense that this is how we normally think of averages. But what's actually going on is we're multiplying each value by a probability, and adding the results. Here, compare the two sides of this equation."

$$\frac{100 + 200 + 300 + 400 + 500 + 600}{6} = 100 \cdot \frac{1}{6} + 200 \cdot \frac{1}{6} + 300 \cdot \frac{1}{6}$$
$$+ 400 \cdot \frac{1}{6} + 500 \cdot \frac{1}{6} + 600 \cdot \frac{1}{6}$$

Average by adding all values and dividing	=	Expected value from multiplying values by probabilities and adding

"On the right side, we're adding up the values that the random variable can take after weighting them according to the probability with which they occur. So we're finding the average of a random variable as a 'sum of probability-weighted values.' Maybe that's clearer as a diagram."

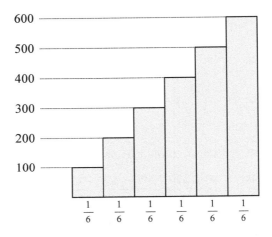

"How do I read this?" Tetra asked.

"Think of the bar heights as showing the random variable values, and their widths as probabilities. If we multiply the heights and widths of each bar and add up the results, what will we get?"

"Um, the total area of all the bars?"

"That's right. But remember, that's also how we found the expected value. So the area of the graph is equivalent to the expected value."

"Okay, sure."

"Check out what happens when we flatten out the bar heights."

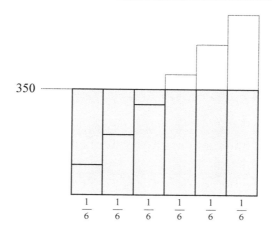

"You've averaged them out."

"Yep. Also, remember that the width of this rectangle is 1, so the area of this figure is the average."

"Huh."

"So anyway, finding expected values is like finding averages. In this case all the probabilities were the same, but we can do the same thing when they aren't. Don't let the sigma freak you out, just be sure you understand what the equation is saying. That an expected value is a sum of probability-weighted values."

c_k ←----→ a specific value that the random variable takes

$\Pr(X = c_k)$ ←----→ probability that the random variable X equals a specific value c_k

$c_k \cdot \Pr(X = c_k)$ ←----→ product of the above two

$\sum_{k=0}^{\infty} c_k \cdot \Pr(X = c_k)$ ←----→ summed values of each $k = 0, 1, 2, 3, \cdots$

"Okay, let me make sure I've got this," Tetra said. She started slowly going through the definition piece by piece, making sure she understood it all. "So in the 100× game, $c_0, c_1, c_2, \ldots, c_5$ would be

$100, 200, 300, \ldots, 600$, right? And $\Pr(X = c_k)$ is the probability of winning c_k yen. That means we get the probability-weighted values by multiplying c_k values by the probabilities that the random variable takes those values, so that's the $c_k \cdot \Pr(X = c_k)$. Then we add it all up, and there's our sum of probability-weighted values."

5.1.7 Fair Game

"So now you know how much you can expect to win in the $100\times$ game, right?" I said.

"Sure, 350 yen!"

"Okay, so what if when you play the game, you have to pay a fee? Let's say that for every roll of the die, you have to pay the house the expected value of your win."

"So I'm paying 350 yen before every roll?"

"Right. If you were to play this game over and over, in the long run you would expect to win 350 yen per roll on average. So there's no real advantage to the house, who's expected to pay out winnings, or to the player, who receives any winnings. That's called a 'fair game.'"

Tetra smiled.

"Thanks for walking me through all that," she said. "There's so much you can learn from a good example that isn't obvious from just equations and definitions."

She let loose a deep sigh just as a strong gust blew in from the window, blowing her notebook off the table.

"Quite a set of lungs you've got there," I joked as I bent over to pick it up.

"It wasn't me! It was the wind!"

"I know, I know. Oh, I've been meaning to ask. What went on with Miruka the other day?"

"What do you mean?"

"She said there was something she wanted to talk to you about?"

"Oh, that. Right. Well, I guess nothing's decided yet, but—"

"Oh, it's decided," came a voice from behind us.

Miruka, right on cue.

5.2 Linearity

5.2.1 Miruka

"Expected values in discrete probability, huh?" Miruka said, peering at Tetra's notebook.

"Yeah, we were just talking about—"

"I'll tell her! Let me!" Tetra said, waving her hands. "I want to see if I can summarize it all."

"Be my guest," I said.

"Okay, we started out talking about random variables. They're functions mapping a sample space to the real numbers, or we can think of them as a named quantity obtained through trials. Then we talked about probability distributions, which represent the probability that a random variable X takes some specific value c. We write them using the probability distribution Pr, as $\Pr(X = c)$. Then we learned about expected values, which are averaged values for a random variable X. We write the expected value of a random variable X as $E[X]$, and it's defined as 'the sum of probability-weighted values.' We can write that as an equation like this."

$$E[X] = \sum_{k=0}^{\infty} c_k \cdot \Pr(X = c_k)$$

Well done, Tetra, I thought to myself. I envied how she never convinced herself that she truly understood a new concept until she had it fully organized in her head. Miruka, however, was less impressed with my teaching skills.

"Why didn't you tell her about the linearity of expectation?" she said, turning to me.

"Linearity—?"

"Of expectation. How the expected value of a sum of random variables is the sum of their expected values."

5.2.2 The Expected Value of a Sum is the Sum of Expected Values

Miruka snagged my notebook and began scribbling equations, Tetra leaning in to watch from one side and me from the other. Such a happy little triangle, I thought. My mother had asked if we had any 'plans,' but I don't think I could have explained to her how *this* was how we enjoyed ourselves together. This is what would tie us together, so long as we had math to share.

Miruka's sharp voice snapped me back from my daydreams.

"Say we have two random variables X and Y, defined in the same sample space. Then $X + Y$, the sum of X and Y, is also a random variable. In this case, we can write this equation using the random variable $X + Y$ and the expected value $E[X + Y]$."

$$E[X + Y] = E[X] + E[Y]$$

"In other words, *the expected value of the sum* of random variables X and Y is equal to *the sum of expected values* X and Y. We can also write this along with some constant K."

$$E[K \cdot X] = K \cdot E[X]$$

"So the expected value of a constant multiple is a constant multiple of the expected value. This means we have two key properties of expected values."

- The expected value of the sum is the sum of expected values.
- The expected value of a constant multiple is a constant multiple of the expected value.

"Together, these properties are called the 'linearity of expectation.' We've only talked about using two variables in these examples, but we can generalize to any number of variables like this. Say we have a random variable X we've created from the sum of n random variables, like this."

$$X = X_1 + X_2 + X_3 + \cdots + X_n$$

"Then we can write an equation like this."

$$E[X] = E[X_1] + E[X_2] + E[X_3] + \cdots + E[X_n]$$

"Excuse me!" Tetra said, her arm shooting straight up.

"Question, Tetra?" Miruka asked.

"Not a question, really, I'd just like to try creating an example of linearity of expectation, if that's okay?"

Well now... Like Miruka, I had thought Tetra had a question, or maybe was going to ask for an example. But taking a shot at making an example on her own, right off the bat? Impressive. "Examples are the key to understanding" had become something of a slogan among us, but Tetra was going beyond just repeating the words; she was putting the idea into practice.

"So let's see," Tetra said. "I guess I should start by making a random variable."

Miruka and I sat quietly as she thought through how to proceed.

"We're talking about sums," Tetra continued, "so I guess I can say we're going to add the results of two die throws. That means we'll need to add the result of the first throw to the result of the second, and I'll call that sum the random variable X. Oh, and I'll make the result of the first throw be the random variable X_1, and the second one X_2. So I guess we have this.

$$X = X_1 + X_2$$

"And what I want to confirm in this example is this, right?"

$$E[X] = E[X_1] + E[X_2]$$

"In other words, I need to make sure that the $E[X]$ on the left is equal to the $E[X_1] + E[X_2]$ on the right!"

Finding $E[X]$ on the left

"The random variable X is the sum of two die throws. That means the values X can possibly have would be from 2, when ⚀⚀ was thrown, to 12, when ⚅⚅ was thrown. So now I need to calculated the expected value of X, in other words $E[X]$. To do that, I guess I need to find the probabilities of $X = 2$, and $X = 3$, and so on up through $X = 12$?"

Tetra cocked her head and glanced at Miruka, who just nodded and said, "Go on."

"Okay," Tetra said. "Then I'll make a table so I can be sure not to make any silly mistakes!"

		2nd throw					
		⚀	⚁	⚂	⚃	⚄	⚅
1st throw	⚀	2	3	④	5	6	7
	⚁	3	④	5	6	7	8
	⚂	④	5	6	7	8	9
	⚃	5	6	7	8	9	10
	⚄	6	7	8	9	10	11
	⚅	7	8	9	10	11	12

Totals when two dice are thrown.

"Each of the $6 \times 6 = 36$ values in this table will occur with probability $\frac{1}{36}$. In other words, we can just count the number of entries in this table to determine probabilities. For example, there are three values of 4 in the table. Here, I'll circle them..."

After doing so, Tetra continued.

"So we can calculate the probability that $X = 4$ like this."

$$\Pr(X = 4) = \frac{3}{36}$$

"Okay, I think we're ready to find $E[X]$ from the definition for expected value. We have to create the sum of probability-weighted values."

$$\begin{aligned}
E[X] &= 2 \cdot \Pr(X=2) + 3 \cdot \Pr(X=3) + 4 \cdot \Pr(X=4) \\
&\quad + 5 \cdot \Pr(X=5) + 6 \cdot \Pr(X=6) + 7 \cdot \Pr(X=7) \\
&\quad + 8 \cdot \Pr(X=8) + 9 \cdot \Pr(X=9) + 10 \cdot \Pr(X=10) \\
&\quad + 11 \cdot \Pr(X=11) + 12 \cdot \Pr(X=12) \\
&= 2 \cdot \frac{1}{36} + 3 \cdot \frac{2}{36} + 4 \cdot \frac{3}{36} + 5 \cdot \frac{4}{36} \\
&\quad + 6 \cdot \frac{5}{36} + 7 \cdot \frac{6}{36} + 8 \cdot \frac{5}{36} \\
&\quad + 9 \cdot \frac{4}{36} + 10 \cdot \frac{3}{36} + 11 \cdot \frac{2}{36} + 12 \cdot \frac{1}{36} \\
&= \frac{2 + 6 + 12 + 20 + 30 + 42 + 40 + 36 + 30 + 22 + 12}{36} \\
&= \frac{252}{36} \\
&= 7
\end{aligned}$$

Tetra looked over what she had written, and nodded.

"There we go. We've found that $E[X] = 7$. In other words, the expected value of the sum is 7."

Finding $E[X_1] + E[X_2]$ on the right

"Okay," Tetra said. "Next we need to find the sum of the expected values. From the definition, the expected value for the first throw, that would be $E[X_1]$, is this."

$$E[X_1] = \sum_{k=1}^{6} k \cdot \text{probability of rolling k}$$
$$= 1 \cdot \frac{1}{6} + 2 \cdot \frac{1}{6} + 3 \cdot \frac{1}{6} + 4 \cdot \frac{1}{6} + 5 \cdot \frac{1}{6} + 6 \cdot \frac{1}{6}$$
$$= \frac{1+2+3+4+5+6}{6}$$
$$= 3.5$$

"So, we have $E[X_1] = 3.5$. That is, the expected value for the first die is 3.5. The calculation for the second throw would be exactly the same, so we can just skip that and say $E[X_2] = 3.5$. When we add those together we get $E[X_1] + E[X_2] = 7$, and we're done! The sum of the expected values is 7. And voila! The expected value of the sum is the sum of expected values!"

$$E[X] = E[X_1] + E[X_2]$$

I noticed that Tetra had become a bit flushed.

"Did I do that right?" she asked.

"You did," Miruka said, short but sweet.

"We can symbolically write 'the expected value of the sum is the sum of expected values' like this," I said, adding a line to the notebook.

$$E\left[\sum (\quad)\right] = \sum \left(E[\quad]\right)$$

"The linearity of expectation means that we can swap out the sum $\sum (\ )$ and the expected value $E[\ ]$," I continued. "Thankfully, the linearity of expectation holds unconditionally for any probability distribution."

5.3 The Binomial Distribution

5.3.1 Speaking of Coins

Miruka stood from her seat and started walking around the table, one finger tracing a circle in the air. I wasn't sure what she was thinking about, but she seemed to be having fun doing it. Each movement of her head cast her hair into waves, which sometimes billowed with the wind entering from the window.

Miruka could be temperamental, quick to anger, even borderline abusive, but her sincere love of mathematics was beyond question. She was also an unfailing supporter of anyone studying math. I still hadn't quite figured out if her personality was incredibly simple or incredibly complex.

"Let's talk about coins," she said, returning to her seat. She pushed her metal-framed glasses higher up on her nose, and started writing in my notebook.

Problem 5-1 (The binomial distribution)

You have a coin with probability p of coming up heads and probability q of coming up tails, where $p + q = 1$. Find the probability $P_n(k)$ that heads comes up k times out of n throws (with $0 \leqslant k \leqslant n$).

"Easy," I said. "If k out of n throws come up heads, then $n - k$ throws came up tails."

"Sure, makes sense," Tetra said, nodding.

"The probability of first throwing k heads, followed by $n - k$ consecutive tails, would be this."

$$\underbrace{\overbrace{\underbrace{p \times p \times p \times \cdots \times p}_{k \text{ values of p}} \times \underbrace{q \times q \times q \times \cdots \times q}_{n - k \text{ values of q}}}^{n \text{ values of p or q}}} = p^k q^{n-k}$$

"Right," Tetra said.

"But of course we don't need all k heads to come first, they just need to come up in the course of the n throws. So that means we need to add up all combinations in which k of the n throws are heads. In other words, we need to multiply this by $\binom{n}{k}$.

"Right!" Tetra agreed. Miruka, however, remained silent as I spoke.

"So the value we're after, $P_n(k)$, would be this."

$$P_n(k) = \binom{n}{k} p^k q^{n-k}$$

Answer 5-1 (The binomial distribution)

$$P_n(k) = \binom{n}{k} p^k q^{n-k}$$

"Good enough," Miruka said. "Even better, $P_n(k)$ is the kth term in the expansion of $(p+q)^n$ by the binomial theorem."

$$(p+q)^n = \binom{n}{0} p^0 q^{n-0} + \binom{n}{1} p^1 q^{n-1} + \binom{n}{2} p^2 q^{n-2} + \cdots$$
$$+ \underbrace{\binom{n}{k} p^k q^{n-k}}_{P_n(k)} + \cdots + \binom{n}{n} p^{n-0} q^0$$

"Oh, neat!" Tetra said.

"Let's write out $(p+q)^n$ as $P_n(k)$."

$$(p+q)^n = P_n(0) + P_n(1) + P_n(2) + \cdots + P_n(k) + \cdots + P_n(n)$$

"Huh, interesting," I said. "And this value equals 1, right? Because $p + q = 1$, so $(p+q)^n = 1$ too."

"How weird," Tetra said. "I thought the binomial theorem was just for expanding two numbers added up and raised to a power, like $(x+y)^n$. Well, I guess it sort of still is, but I never would have thought it's also related to probability distributions for how many times you get heads when you flip a coin. Er, hang on... $P_n(k)$ is a probability distribution, right?"

"It is if you think of k as a variable," Miruka said. "It's a probability if you think of k as a constant. And if you think of $P_n(k)$

as a probability distribution, we call that a binomial distribution. We're divvying up a 1 among $P_n(0), P_n(1), P_n(2), \ldots, P_n(n)$. If you flip a coin n times and represent the number of heads as a random variable X, then X follows a binomial distribution."

Miruka paused, thought, and held a finger in the air.

"Okay," she said, "let's find the expected value of a random variable X that follows the binomial distribution."

5.3.2 Expected Value of a Binomial Distribution

Problem 5-2 (Expected value of a binomial distribution)

You have a coin with probability p of coming up heads and probability q of coming up tails, where $p + q = 1$. Find the expected value for the number of heads.

Miruka pointed at Tetra with a flourish.

"What's the answer when $n = 3$?" she said.

"Oh, okay. Let me find $E[X]$," Tetra said, reaching for her notebook and mechanical pencil.

"Wait," Miruka said, stopping her. "You have to say what X is going to represent first."

"Oh, right. We have to be sure to bring in our random variable. I guess X should be the number of times we get heads. So what I'm looking for is the expected value of X when $n = 3$."

"Much better."

"Great! Okay, so from the definition of expected value for a random variable, this should be true."

$$E[X] = \sum_{k=0}^{\infty} k \cdot \Pr(X = k)$$

"From the definition of the binomial theorem, I can rewrite this $\Pr(X = k)$ part."

$$= \sum_{k=0}^{\infty} k \cdot \binom{n}{k} p^k (1-p)^{n-k}$$

"If $k > n$, then $\binom{n}{k} = 0$, so I just need to think about the sum up until n"

$$= \sum_{k=0}^{n} k \cdot \binom{n}{k} p^k (1-p)^{n-k}$$

"Okay, so here I want $n = 3$."

$$= \sum_{k=0}^{3} k \cdot \binom{3}{k} p^k (1-p)^{3-k}$$

"The $k = 0$ term will just be 0, so I just need to consider $k = 1, 2, 3$."

$$= \sum_{k=1}^{3} k \cdot \binom{3}{k} p^k (1-p)^{3-k}$$

"Now I'll expand the sigma."

$$= 1 \cdot \binom{3}{1} p^1 (1-p)^2 + 2 \cdot \binom{3}{2} p^2 (1-p)^1 + 3 \cdot \binom{3}{3} p^3 (1-p)^0$$

"Um, here I guess I use $\binom{3}{1} = 3$, $\binom{3}{2} = 3$, $\binom{3}{3} = 1$."

$$= 1 \cdot 3p^1 (1-p)^2 + 2 \cdot 3p^2 (1-p)^1 + 3 \cdot 1p^3 (1-p)^0$$

"Cleaning up ... "

$$= 3p(1-p)^2 + 6p^2(1-p) + 3p^3$$

"... and expanding $(1-p)^2$ and $p^2(1-p)$."

$$= 3p(1 - 2p + p^2) + 6(p^2 - p^3) + 3p^3$$

"A little more expanding here."

$$= 3p - 6p^2 + 3p^3 + 6p^2 - 6p^3 + 3p^3$$

"Then combining like terms, I get ... "

$$= 3p + (6-6)p^2 + (3-6+3)p^3$$

"Huh?"

$$= 3p$$

"What happened?" Tetra said. "All the p^2 and p^3 terms disappeared! I ended up with just 3p!"

$E[X] = 3p$ Expected value for random variable X following binomial distribution $P_3(k)$

"Interesting," I said. "If $E[X] = 3p$ when $n = 3$, then..."

"I know! I know!" Tetra said. "The general form would be $E[X] = np$, right? Sure it would! Here, let me show it. I can use mathematical induction for n!"

Tetra was really on a roll today. I could hear her breathing faster as she snatched up her pencil again.

"Wait," Miruka said, stopping her again. "Let's do it using linearity of expectation."

5.3.3 Separation into Sums

"Using linearity of expectation?" Tetra said. "Um, how?"

"Linearity of expectation suggests we should separate random variables into sums," Miruka said.

"Um..."

"We're trying to find the expected value for the number of heads in n coin flips. What's the random variable we should be focusing on?"

"That would be the number of heads. We represent that using the random variable X."

"Well then, let's consider a new random variable X_k," Miruka said. "We're going to say that X_k is 1 if the kth coin flip is heads, and 0 if it's tails."

- X_k is 1 if the kth coin flip is is heads

- X_k is 0 if the kth coin flip is is tails

"One for heads, zero for tails..." Tetra repeated, jotting this down in her notebook.

"Oh, it's an indicator!" I said. "Like we talked about for the linear search algorithm. We used a variable S that was 1 when we had found something, and 0 when we didn't."

- S is 1 if we have found something

- S is 0 if we have not found something

"Very good," Miruka said, smiling. "We'll call a variable like X_k, one that equals 1 when some event occurs and 0 when it doesn't, an indicator random variable. Plenty of other things you can call it though—an indicator variable, a dummy variable, a Bernoulli random variable, a 0–1 random variable..."

"Interesting," I said.

"Uh, sorry. What's this k in the X_k?" Tetra asked.

"It's some number from 1 to n," Miruka said. "If we take the sum of all the X_k indicator random variables, that sum should equal the random variable X. Do you see why, Tetra?"

$$X = X_1 + X_2 + X_3 + \cdots + X_k + \cdots + X_n$$

Tetra shook her head. "No, no I do not."

"Keep on following what the random variable represents," Miruka said. "X is a random variable representing the number of heads. X_k is a random variable that takes a value of 1 if the kth coin toss was heads, and 0 if it was tails."

"Yeah, sure..."

"So what happens if $n = 3$, and the flips were heads, tails, heads?"

"Oh, right, an example. Let me write this out."

- Coin toss #1: heads (so $X_1 = 1$)
- Coin toss #2: heads (so $X_2 = 0$)
- Coin toss #3: heads (so $X_3 = 1$)

"There were two heads," Tetra said, "so $X = 2$. Oh, I get it! So $X = X_1 + X_2 + X_3$ holds! X_k tracks whether the kth flip was heads, in which case we assign it a 1, or tails, in which case we assign it a 0. When we add up all the X_k's, we get X, in other words the total number of heads! Of course!"

"Well done," Miruka said, nodding.

5.3.4 Indicator Random Variables

"Indicator random variables are convenient for counting," Miruka continued. "As an example, let's consider an indicator random variable C, which will take a value of 1 if a single coin toss comes up heads, or 0 if it comes up tails. Given that, what is the expected value $E[C]$ for the random variable C? We can calculate that, keeping in mind that C can only take one of two values."

$$E[C] = 1 \cdot \Pr(C = 1) + 0 \cdot \Pr(C = 0)$$
$$= \Pr(C = 1)$$

"In other words..."

$$E[C] = \Pr(C = 1)$$

"This equation tells us that the expected value of an indicator random variable is the same value as the probability that the indicator random variable equals 1."

Miruka paused to let us take that in, then continued.

"Back to the problem at hand," she said, "finding the expected value $E[X]$, which represents the number of heads appearing in n coin tosses. We start by breaking X up into a sum."

$$E[X] = E[X_1 + X_2 + X_3 + \cdots + X_n]$$

"Use the linearity of expectation."

$$= E[X_1] + E[X_2] + E[X_3] + \cdots + E[X_n]$$

"X_k is an indicator random variable, so its expected value equals the probability."

$$= \Pr(X_1 = 1) + \Pr(X_2 = 1) + \Pr(X_3 = 1) + \cdots + \Pr(X_n = 1)$$

"From the problem, the probability that the kth toss comes up heads is p."

$$= \underbrace{p + p + p + \cdots + p}_{n \text{ of these}}$$
$$= np$$

> **Answer 5-2 (Expected value of the binomial distribution)**
>
> For a flipped coin with probability p of coming up heads and probability q of coming up tails, the expected value for the number of heads is
>
> $$np.$$

"Whoa," Tetra said. "That feels...sudden."

"Only because we have an indicator random variable ticking off the number of heads," Miruka said.

- Linearity of expectation suggests that we should <u>break the random variable up into a sum</u>.

- Indicator random variables suggest that we should <u>find expected values as probabilities</u>.

"When we combine these two, finding expected values is a piece of cake."

5.3.5 Some Fun Homework

"The library is *closed!*" came Ms. Mizutani's harsh announcement.

Already?

"Um, Miruka?" Tetra asked as we were cleaning up. "Why are we talking about averages and expected values and all that?"

"Because we want to think about things quantitatively," Miruka replied. "A random variable can take a number of values, depending on the trial. When we have to handle lots of values, it's natural to want to summarize them. An averaged value, in other words an expected value, is one way of summarizing the values that a random variable can take."

"A summarized value...?"

"Here," Miruka said, handing me a card. "Some fun homework."

> **Problem 5-3 (Expected value until all faces come up)**
>
> Find the expected value for the number of throws of a die thrown until all faces come up at least once.

"Is this from Mr. Muraki?" I asked.

"It is," Miruka said, "but I've already done it. I love coin-flipping, you know."

I decided it was probably best not to correct her slip, saying "coin-flipping" in place of "dice-throwing."

5.4 Everything Happening

5.4.1 Someday

Back home, alone in my room, late at night with my parents already asleep and my school homework finished: the perfect time for me to work on my own math.

But I found myself thinking about Tetra. About how she didn't just say "I don't understand," she followed that up with thinking about just *what* she didn't understand. About how she didn't just ask questions, she created her own examples. About how she didn't just listen to explanations, she summarized the key points of what she heard. I was amazed at how much she had grown since we'd first met, which made me all the more aware of how much growing I still needed to do myself.

My thoughts then turned to Miruka, how she had jumped straight from expected values to linearity of expectation. Easy for her, no doubt, because the two were bound together, mingling with all those other concepts that made up the beautiful mathematical cosmos within her head. Linearity of expectation is obvious... *after* it has been explained to you. It wasn't, however, something that I could have come up with on my own. In that way, Miruka had me beat too.

Alright, that's enough self-deprecation for one night.

There was no point in comparing my own abilities with those of others. Pretty much every problem I had ever worked on had already been solved by somebody, somewhere, so from an objective point of view I wasn't accomplishing much by solving them on my own. Subjectively, however, working through these problems on my own was very meaningful. Even the ones I couldn't solve.

Someday, I would have the skills to take on problems that no one had solved before. Someday *I* would be the one sending messages to future mathematicians.

5.4.2 Throwing Dice

> **Problem 5-3 (Expected value until all faces come up)**
>
> Find the expected value for the <u>number of throws of a die</u> thrown until all faces come up at least once.

The problem didn't look all that hard on the surface, but I'd been fooled before. I decided to take the safe route, and start with an example. I grabbed a pencil and notebook to write out my thoughts.

So I'm throwing a die... Saying that every face comes up with equal probability seemed like a safe assumption. Also, a die has six faces, from ⚀ to ⚅, and I need to keep throwing until all six faces show up at least once. *Right, got it.*

The extreme example would be where I made six throws, and rolled ⚀ to ⚅ in turn.

⚀ → ⚁ → ⚂ → ⚃ → ⚄ → ⚅   All six faces in 6 throws

But of course all I needed to do is get all six faces among the six throws; the order wouldn't matter.

⚂ → ⚀ → ⚄ → ⚅ → ⚁ → ⚅   All six faces in 6 throws

Not only that, but I would have to be incredibly lucky to get all six faces in just six throws. For example, I might throw a second ⚀, and therefore need seven throws to get all six faces.

⚂ → ⚀ → ⚄ → <u>⚀</u> → ⚄ → ⚁ → ⚅   All six faces in 7 throws

Well, then. The end goal was to find the number of die throws until all faces have appeared, so I decided to use a random variable to keep track of that.

Let random variable X be the number of die throws until all faces appear

If I was really lucky, then X would be 6. In most cases, however, it would be much larger. Even if I started out with good throws, it might take a while before the last one I needed showed up.

$$\boxed{\cdot\cdot} \to \boxed{\cdot} \to \boxed{\cdot\cdot} \to \boxed{\cdot} \to \boxed{\cdot\cdot} \to \boxed{::} \to$$

$$\boxed{::} \to \boxed{\cdot\cdot} \to \boxed{\cdot\cdot} \to \boxed{\cdot\cdot} \to \boxed{\cdot\cdot} \to \boxed{::} \to \boxed{\cdot}$$

All six faces finally show up after 13 throws

By this point I was pretty sure I had a good grip on the problem. I had set the number of throws needed before all six faces appeared as the random variable X, so the solution to this problem would be the expected value of X, in other words E[X]. The definition of expected value was $E[X] = \sum_{k=0}^{\infty} c_k \cdot \Pr(X = c_k)$, so what I needed to calculate was $\Pr(X = c_k)$. I knew a few things about this value. For example, the probability of X = 1 would be 0, since there's no way to get all six faces in just one throw, so $\Pr(X = 1) = 0$. In fact, I could say the same thing about all values of X less than 6; $\Pr(X = 2), \Pr(X = 3), \Pr(X = 4), \Pr(X = 5)$ would all be 0.

Well then, what about $\Pr(X = 6)$? Getting all six faces in exactly six throws would mean no duplicate throws, and I knew how to calculate that.

- It wouldn't matter what the 1st throw was, so there are 6 possibilities. For each of those...

- On the 2nd throw I want any other face, so there are 5 possibilities. For each of those...

- On the 3rd throw I want any face that hasn't shown up yet, so there are 4 possibilities. For each of those...

- On the 4th throw I want any face that hasn't shown up yet, so there are 3 possibilities. For each of those...

- On the 5th throw I want any face that hasn't shown up yet, so there are 2 possibilities. For each of those...

- On the 6th throw I want any face that hasn't shown up yet, so there is only 1 possibility.

So here's what I've got.

$$\Pr(X = 6) = \frac{6 \times 5 \times 4 \times 3 \times 2 \times 1}{6 \times 6 \times 6 \times 6 \times 6 \times 6}$$

$$= \frac{6!}{6^6}$$

Next, I considered the case where $\Pr(X = 7)$, one throw repeating a face that had already been thrown. There were six possibilities for the doubling, 1 through 6, and I also had to account for where in the seven throws the doubling occurred.

For example, I could consider the case where the ⚅ repeated somewhere in ⚀ → ⚁ → ⚂ → ⚃ → ⚄ → ⚅.

⚀ → ⚅ → ⚁ → ⚂ → ⚃ → ⚄ → ⚅
⚀ → ⚁ → ⚅ → ⚂ → ⚃ → ⚄ → ⚅
⚀ → ⚁ → ⚂ → ⚅ → ⚃ → ⚄ → ⚅
⚀ → ⚁ → ⚂ → ⚃ → ⚅ → ⚄ → ⚅
⚀ → ⚁ → ⚂ → ⚃ → ⚄ → ⚅ → ⚅ ... huh?

Hang on. This won't do. My examples have been wrong from the start! For example, what if the seven throws were this?

⚀ → ⚅ → ⚁ → ⚂ → ⚃ → ⚄ → ⚅

In this case, the sixth face didn't show up on the seventh throw, that happened on the sixth, when I threw a ⚅. So here I had $X = 6$, not $X = 7$.

This is where I realized that sure enough, the problem wasn't going to be as easy as it had seemed at first. And things could only get worse for $\Pr(X = 8)$! If I kept on like this, there was no way I could find $\Pr(X = c_k)$ for an arbitrary k. And if I couldn't do that, I couldn't calculate the expected value $\mathsf{E}[X]$.

I stayed up later than I should have, trying a little of this and a little of that, but no clear solution revealed itself. Just about the time I had resigned myself to having to work through a small mountain of calculations, sleep overcame me. I had a strange dream about trying to roll dice, but each time I did they turned into coins.

"I need dice, not coins!" my dream self complained.

"These dice *are* coins," dream Miruka chided.

5.4.3 The Stairway of Happiness

"Good mo—rning!" Tetra sang.

"Morning," I mumbled.

Tetra had joined me on the way to school, forcing me to walk faster to keep up.

"Did you solve Miruka's problem?" she asked.

"I wandered into a maze of calculations, and got lost," I admitted.

Tetra shook her head. "I didn't even get as far as the calculations," she said. "I still have no idea how to break everything up into a sum."

"How to... what?" I said, stopping. Tetra too stopped next to me.

"Huh? What's wrong?"

"You said something about breaking everything up into a sum."

"Sure. We have to use that thing about the expected value of the sum being the sum of the expected values, right?"

Linearity of expectation. I am so, so stupid...

Miruka had spent so much time talking about it, practically chanted it. Tetra had even written out an example using it! But I had wasted so much time trying to approach the problem directly, from the definition of expected values. I hadn't even *tried* to break up my random variable X into a sum.

'Stupid' doesn't even begin to—

"You okay?" Tetra said, looking up at me with worry in her eyes.

"Sure, sorry. Nothing to do with you. I was just shocked at how much of a fool I could be." I took a deep breath. "I had completely forgotten about linearity of expectation. How far did it get you?"

Tetra looked relieved, and we resumed walking.

"Not far," she said. "I just figured that to use linearity of expectation we had to break the number of die throws into a sum somehow. So I made a lot of examples, but I wasn't really sure what to do with them. I did come up with the stairway of happiness, though!"

"The stairway of... happiness?"

"Hang on a sec." Tetra stopped again and pulled a sheet of paper out of her bag.

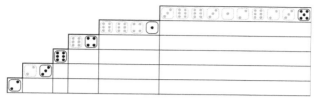

A 'stairway of happiness' for die throws.

I could feel my pulse quickening, intuitively knowing that something important was going on in this figure.

"What am I looking at here?" I asked.

"Well, you start from the left, and move to the right," Tetra said, pointing at the columns in turn. "Start with the the first step. You roll a ⚀, and you go up one step. At the next step you get another ⚀, so you stay where you are, but then you roll ⚁ and you get to go up one more step. At the third step you get ⚃, and you can go up again. In other words, you stay on a step until you roll a number you haven't rolled before. When you roll a face you haven't seen yet, you get to climb one more step up the staircase."

"Huh, interesting."

"So the steps stay flat while you're rolling numbers you've seen, and jump up when you roll something new."

"And why is it the 'stairway of happiness'?"

"I dunno. I just figured the dice would be happier the higher up they got."

"Uh, right."

"But I found something interesting! I pulled some numbers out of the square root of five to use as the sequence of rolls, but can you believe that the first 5 doesn't show up until the thirty-sixth digit after the decimal? Isn't that *weird*?"

$$\sqrt{5} = 2.\underline{23}6\underline{0}\underline{6797}\underline{4}99789\underline{6}\underline{96}\underline{40}\underline{91}\underline{73668}\underline{731276235}\cdots$$

With apologies to Tetra, I already wasn't paying much attention to what she was saying.

"You've already found it," I said.

"Found what?" she asked, blinking rapidly.

"How to break the number of throws up into a sum."

"Er, I did?"

"It's right there, in your stairway of happiness. The total length of the stairway is the number of die throws you need to see all the faces. And the total length of the stairway is the sum of the length of each step!"

"Oh. Oh!"

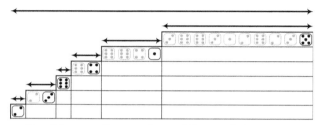

Breaking up the number of throws into a sum.

5.4.4 Measuring the Staircase

After school I found myself in our classroom, standing behind the podium. Tetra and Miruka sat in the front row of desks. I was preparing to explain to Miruka how Tetra's "stairway of happiness" had led us to a solution to the problem, finding the expected value for the number of die throws before all faces come up.

"Let random variable X be the number of dice thrown before all faces come up at least once," I began. "This X corresponds to the overall width of Tetra's stairway of happiness. We're also going to use a random variable X_j in a somewhat awkward way, to represent 'the number of dice throws until some previously unseen face appears after j different faces have come up.' In other words, X_j is the length of step $j+1$ in the stairway."

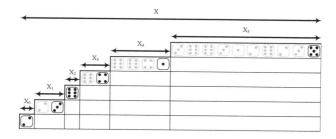

"So X is the sum of the X_j values, and in this case we can think of j taking values from 0 to 5."

$$X = X_0 + X_1 + X_2 + X_3 + X_4 + X_5$$

"We want to find the expected value $E[X]$, so we'll use the fact that the expected value of the sum is the sum of the expected values."

$$E[X] = E[X_0 + X_1 + X_2 + X_3 + X_4 + X_5]$$
$$= E[X_0] + E[X_1] + E[X_2] + E[X_3] + E[X_4] + E[X_5]$$

"So let's consider the random variable X_j. At the start we haven't thrown any dice yet, so no faces have shown up. That means when we throw the first die we will definitely see a face that hasn't come up yet, which we write like this."

$$X_0 = 1$$

"For every dice throw after this, there are two possibilities."

· The face that comes up is a new one.

· The face that comes up is one we've already seen.

"What we want to know is, when we've already seen j different faces, what is the probability that some not-yet-seen face will come up? If we've already seen j faces, that means there are $6 - j$ unseen faces remaining. So we can let p_j be the probability of an unseen face being rolled, and calculate it like this."

$$p_j = \frac{6-j}{6} = 1 - \frac{j}{6}$$

"We can also calculate the probability of rolling a face we've already seen, given that we've already seen j faces, which we'll call q_j. Since there are six faces, and we've seen j of them, we calculate q_j like this."

$$q_j = \frac{j}{6}$$

"We talk about these probabilities p_j and q_j while we're on step $j+1$ of the stairway of happiness. Once we've rolled a face we haven't seen yet, we go up a step and talk about probabilities p_{j+1} and q_{j+1}."

Miruka snapped her fingers to get my attention.

"In other words," she said, "we're flipping coins with varying probabilities."

I gulped. "So *that's* what you meant!"

"What's what she meant?" Tetra said, still scribbling in her notebook.

"She means we can forget about the dice in this problem! We just have to think of tossing a coin with probability p_j of coming up heads and probability q_j of coming up tails. We go up a step each time we throw a heads, but every time that happens the probability of getting heads decreases. The coin looks like this."

- The probability of flipping heads is $p_j = 1 - \frac{j}{6}$.
- The probability of flipping tails is $q_j = \frac{j}{6}$.

"Now that we understand the nature of the stairway of happiness, let's think about random variable X_j, which is the length of step $j+1$. We want to know the probability of that length becoming k, in other words the probability $\Pr(X_j = k)$. This will be equal to the probability of flipping heads after flipping $k-1$ tails in a row. This is what makes Miruka's coin-flipping model so much easier to explain. Like this."

$$\begin{aligned}
\Pr(X_j = k) &= q_j^{k-1} \cdot p_j && \text{probability of flipping } k-1 \text{ tails,} \\
& && \text{then heads} \\
&= q_j^{k-1} \cdot (1 - q_j) && \text{use } p_j = 1 - q_j \\
&= q_j^{k-1} - q_j^k && \text{expand}
\end{aligned}$$

"From this we get that $\Pr(X_j = k)$, where $j = 0, 1, 2, 3, 4, 5$ and $k = 1, 2, 3, \ldots$, and this will let us calculate the expected value of the random variable X_j, in other words $E[X_j]$. For some appropriate value of n, we want to find the partial sum $\sum_{k=1}^{n} k \cdot \Pr(X_j = k)$, and find the limit when n goes to infinity."

$$\sum_{k=1}^{n} k \cdot \Pr(X_j = k) = 1 \cdot \Pr(X_j = 1)$$
$$+ 2 \cdot \Pr(X_j = 2)$$
$$+ 3 \cdot \Pr(X_j = 3)$$
$$+ \cdots$$
$$+ n \cdot \Pr(X_j = n)$$
$$= 1 \cdot (q_j^0 - q_j^1)$$
$$+ 2 \cdot (q_j^1 - q_j^2)$$
$$+ 3 \cdot (q_j^2 - q_j^3)$$
$$+ \cdots$$
$$+ n \cdot (q_j^{n-1} - q_j^n)$$
$$= 1 \cdot q_j^0 - 1 \cdot q_j^1$$
$$+ 2 \cdot q_j^1 - 2 \cdot q_j^2$$
$$+ 3 \cdot q_j^2 - 3 \cdot q_j^3$$
$$+ \cdots$$
$$+ n \cdot q_j^{n-1} - n \cdot q_j^n$$
$$= q_j^0 + q_j^1 + q_j^2 + q_j^3 + \cdots + q_j^{n-1} - n \cdot q_j^n$$

"We can calculate this as the sum of a geometric progression."

$$= \frac{1 - q_j^n}{1 - q_j} - n \cdot q_j^n$$

"Now we just need to find the limit. Since $q_j = \frac{j}{6}$, we know that

"$0 \leqslant q_j < 1$, so the limit will converge."

$$\begin{aligned}
E[X_j] &= 1 \cdot \Pr(X_j = 1) \\
&\quad + 2 \cdot \Pr(X_j = 2) \\
&\quad + 3 \cdot \Pr(X_j = 3) \\
&\quad + \cdots \\
&\quad + k \cdot \Pr(X_j = k) \\
&\quad + \cdots \\
&= \lim_{n \to \infty} \sum_{k=1}^{n} k \cdot \Pr(X_j = k) \\
&= \lim_{n \to \infty} \left(\frac{1 - q_j^n}{1 - q_j} - n \cdot q_j^n \right) \\
&= \frac{1}{1 - q_j} \\
&= \frac{1}{1 - \frac{j}{6}} \quad \text{rewrite } q_j \text{ as } \frac{j}{6} \\
&= \frac{6}{6 - j}
\end{aligned}$$

"In other words, the expected value for step $j + 1$ is this."

$$E[X_j] = \frac{6}{6-j}$$

"And now, finally, we can find the expected value for the length of the entire staircase."

$$\begin{aligned}
E[X] &= E[X_0 + X_1 + X_2 + X_3 + X_4 + X_5] \\
&= E[X_0] + E[X_1] + E[X_2] + E[X_3] + E[X_4] + E[X_5] \\
&= \frac{6}{6-0} + \frac{6}{6-1} + \frac{6}{6-2} + \frac{6}{6-3} + \frac{6}{6-4} + \frac{6}{6-5} \\
&= \frac{6}{6} + \frac{6}{5} + \frac{6}{4} + \frac{6}{3} + \frac{6}{2} + \frac{6}{1} \\
&= 6 \cdot \left(\frac{1}{6} + \frac{1}{5} + \frac{1}{4} + \frac{1}{3} + \frac{1}{2} + \frac{1}{1} \right) \\
&= 6 \cdot \left(\frac{1}{1} + \frac{1}{2} + \frac{1}{3} + \frac{1}{4} + \frac{1}{5} + \frac{1}{6} \right)
\end{aligned}$$

"So if we want to roll a die until every face comes up at least once, then the expected number of rolls we'll have to make is this."

$$E[X] = 6 \cdot \left(\frac{1}{1} + \frac{1}{2} + \frac{1}{3} + \frac{1}{4} + \frac{1}{5} + \frac{1}{6} \right)$$

"Wow, things sure do clean up nice," I said.

"Very nicely indeed," Miruka said, a satisfied expression on her face.

Tetra looked up from her notebook. "I just calculated it!" she said. "$E[X] = 14.7$, so on average we would expect to have to roll the die 14.7 times! That's more than I would have expected!"

Answer 5-3 (Expected value until all faces come up)

We can expect to roll the die

$$6 \cdot \left(\frac{1}{1} + \frac{1}{2} + \frac{1}{3} + \frac{1}{4} + \frac{1}{5} + \frac{1}{6} \right) = 14.7$$

times.[a]

---
[a]Problem 5-3 is a variation on a classic problem called "the coupon collector's problem." Strictly speaking, the formulation requires that the sample space be an infinite set.

"Let's write it as a harmonic number," Miruka said.

"Um, remind me what a harmonic number is?" Tetra said.

Miruka wrote an equation in my notebook, and showed it to Tetra. I left the podium and went to Tetra's desk to take a look.

$$H_n = \frac{1}{1} + \frac{1}{2} + \frac{1}{3} + \cdots + \frac{1}{n}$$

"If we use H_n, we can write the expected value of X like this."

$$E[X] = 6 \cdot H_6$$

"We can even generalize the problem, because why limit ourselves to six-sided dice? Repeating the same calculation gives us this."

$$E[X] = n \cdot H_n$$

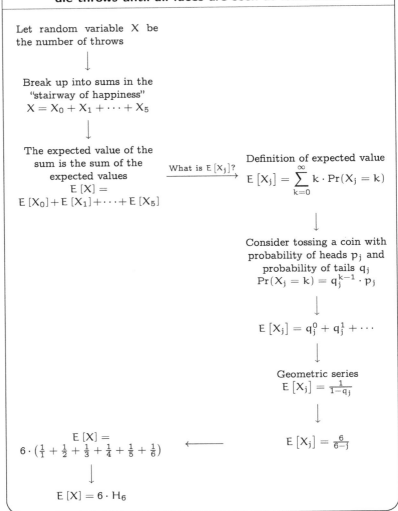

5.4.5 Slip of the Tongue

"It's so much fun working through problems like this together!" Tetra said.

"I was only able to come up with an answer thanks to your 'stairway of happiness,'" I said.

"I just wish my stairway had led me to a solution too," she said. She was smiling, though, and I returned her smile.

Miruka was smiling too, apparently caught up in the mood. Perhaps more than usual, because she even started to give me some uncharacteristic praise.

"Good job, —" she started.

But then...

Then...

Instead of my name, she called me by her brother's name.

Time froze. Miruka, Tetra, and I sat motionless, in stunned silence.

Everyone—even the normally perfect Miruka—makes a slip of the tongue from time to time, even calls someone by the wrong name.

But this... calling me by the name of her dead brother...

Miruka shattered the frozen tableau by hurling my notebook at my face and stalking out through the door.

> If we want to understand the typical behavior of a given random variable, we often ask about its "average" value.
>
> GRAHAM, KNUTH, & PATASHNIK
> *Concrete Mathematics* [8]

My notes

The binomial distribution and sample spaces

In problem 5-2 (expected value of a binomial distribution), we found the expected number of heads among n coin flips. When doing so we considered a sum of random variables, but how should we consider the sample space behind all that?

If we consider n throws as one trial, then we can think of a sample space Ω as follows.

$$\Omega = \{\langle u_1, u_2, ..., u_n \rangle \mid u_k \in \{H, T\}, 1 \leq k \leq n\}$$

So if $n = 3$, the sample space Ω would be this.

$$\begin{aligned}\Omega &= \{\langle u_1, u_2, ..., u_n \rangle \mid u_k \in \{H, T\}, 1 \leq k \leq 3\} \\ &= \{\langle H, H, H \rangle, \langle H, H, T \rangle, \langle H, T, H \rangle, \langle H, T, T \rangle, \\ &\quad \langle T, H, H \rangle, \langle T, H, T \rangle, \langle T, H, H \rangle, \langle T, T, T \rangle\}\end{aligned}$$

We can let X be a random variable representing the number of heads, and let X_k be an indicator random variable taking a value of 1 if throw k is heads or 0 if throw k is tails.

ω	$X(\omega)$	$X_1(\omega)$	$X_2(\omega)$	$X_3(\omega)$
$\langle H, H, H \rangle$	3	1	1	1
$\langle H, H, T \rangle$	2	1	1	0
$\langle H, T, H \rangle$	2	1	0	1
$\langle H, T, T \rangle$	1	1	0	0
$\langle T, H, H \rangle$	2	0	1	1
$\langle T, H, T \rangle$	1	0	1	0
$\langle T, T, H \rangle$	1	0	0	1
$\langle T, T, T \rangle$	0	0	0	0

When we look at this table, it's clear that

$$X(\omega) = X_1(\omega) + X_2(\omega) + X_3(\omega)$$

for any value of $\omega \in \Omega$

Chapter 6

Indiscernible Future

> And it was after long searching that I found out the carpenter's chest, which was, indeed, a very useful prize to me, and much more valuable than a shipload of gold would have been at that time.
>
> DANIEL DEFOE
> *Robinson Crusoe*

6.1 MEMORY OF A PROMISE

6.1.1 *On a Riverbank*

"We'll do more math tomorrow, he said."

I was on a riverbank, sitting next to Miruka and watching the sky slowly redden toward sunset. Two crows flew by. I could hear a train, somewhere far off. There was a little wind, but it wasn't cold. No one else could be seen.

"Tomorrow," she repeated. I came close to asking who had said that, but I held my tongue. Of course I knew. "He sent me home from the hospital. He said we could do more the next day."

Her voice was different. Softer. Almost childish.

"I didn't want to leave, but he promised we could work on more math the next day. He *promised*."

She leaned in toward me, and I put an arm around her. She placed her head on my shoulder, and I smelled oranges.

So warm...

We sat there silently for a while, at the place where I had finally caught up with her after she had run from the classroom. My nose still throbbed a bit from where she had hit me with my notebook, but the pain was already receding. I stole a glance, and saw that her eyes were closed.

I found myself thinking about what was right, what it meant to be right about something. I was rarely sure about being right outside of mathematics, but this was an exception—it felt right to be sitting there beside Miruka.

The sky edged toward crimson, promising darkness soon. Miruka gave a loud sigh, stood up, and brushed the grass off of her skirt. I stood too, and turned to face her. I reached out and ran a finger down her cheek, tracing the path of a tear. Miruka grasped my hand with surprising strength.

"I never should have left him," she said.

6.2 Order

6.2.1 Fast Algorithms

Several days later it was Golden Week, Japan's nearly weeklong series of spring holidays, but there was no rest for those of us facing college entrance exams. I had spent the morning taking a special test-prep seminar, then went to the library to work on some review problems. After working through a few I paused to take a break, and noticed Tetra and Lisa at another table, huddled in an intense conversation. Well, to be more accurate, Tetra was the intense one. Lisa's input was mostly nods and shakes of her head as she silently typed on her flame-red computer.

"Oh, hey!" Tetra said, waving.

"Studying more algorithms?" I asked, approaching them.

"Not algorithms per se," Tetra said, looking at her notes. "Algorithm speed. Remember the other day, when we analyzed the linear search algorithm and its improved version, the sentinel linear search? We found that both algorithms took the longest time when the result was ⟨NOT FOUND⟩, and that the maximum number of steps was like this."

Algorithm	Max. steps
LINEAR-SEARCH	$T_L(n) = 4n + 5$
SENTINEL-LINEAR-SEARCH	$T_S(n) = 3n + 7$

Maximum number of steps for an array of size n.

"Sure, I remember this," I said, looking at her notes. "But what's this $T_L(n)$?"

"Oh, things were getting messy, so we decided to give names to the maximum number of steps. We're using $T_L(n)$ for LINEAR-SEARCH, and $T_S(n)$ for SENTINEL-LINEAR-SEARCH."

"I suppose the n is the array size?"

"That's right! And we're comparing their speeds."

"When you compare two mathematical statements you often use subtraction," I said. "Checking to see whether the difference between two things is positive or negative is the very basis of comparisons like that. Like, if n is a natural number we can do this."

$$\begin{aligned} T_L(n) - T_S(n) &= (4n + 5) - (3n + 7) \\ &= 4n - 3n + 5 - 7 \\ &= n - 2 \end{aligned}$$

"This means that $T_L(n) - T_S(n) > 0$ only when $n > 2$. We can also move the $T_S(n)$ to the right side to state it as an inequality."

$$T_L(n) > T_S(n) \qquad (\text{for } n > 2)$$

"Sure, that's easy enough. This means the sentinel linear search algorithm will complete with fewer steps—in other words, faster—for any array with more than two elements. But when we talked about this before, Miruka also said something about, what was it..." Tetra placed a finger on her lips as she thought, "Oh! Asymptotic analysis! How that was important somehow. I was just wondering if that was some way of going deeper than making sure that an inequality like this holds. Do you know?"

"No idea," I admitted. "But I think you're right. Here, look at this graph."

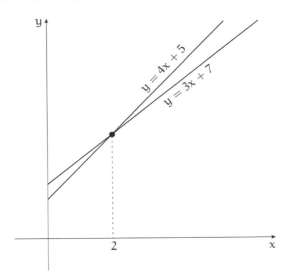

"See? When $n > 2$, the graph of $y = 3x + 7$ is always below $y = 4x + 5$. That's another way to see that the sentinel linear search will always complete with fewer steps than the linear search under that condition."

"Yeah, the graph does make that a lot clearer, doesn't it."

"Gyaak!" Lisa shouted. I turned to see her batting away Miruka, who had sneaked up behind us and was amusing herself by playing with Lisa's hair.

6.2.2 Of Order At Most n

"Uh, Miruka?" Tetra said. "Can you tell us if there's a more precise way of analyzing algorithm times like $T_L(n) = 4n + 5$ and $T_S(n) = 3n + 7$?"

A serene Miruka scanned our faces. There was no trace of the girl I had seen on the riverbank just a few days ago.

"Don't assume that analysis leads to precision," she said. "If you're interested in analysis of algorithms, we need to talk about big-O notation[1]."

Miruka made a "gimme" gesture, and I handed her a notebook and pencil.

---

[1]"Big-O notation" is also referred to as "$O(n)$," "order n," etc.

"You can describe the behavior of a function $T(n)$ using big-O notation like this."

$$T(n) = O(n)$$

"What we're doing here is showing how quickly the value of function $T(n)$ increases as n increases. Let's consider $T(n)$ as representing the number of steps the algorithm executes, which you can also think of as the execution speed. Big-O notation is a quantitative representation of how much longer it take the algorithm to complete when the size of its input n gets larger. This statement $T(n) = O(n)$ says that there is a natural number N and a positive number C so that this inequality will hold for any integer n equal to or larger than N."

$$|T(n)| \leqslant Cn$$

"If you want to be a stickler and define the entire thing as a logic statement, use this."

$$\exists N \in \mathbb{N} \; \exists C > 0 \; \forall n \geqslant N \; \Big[\, |T(n)| \leqslant Cn \,\Big]$$

"In this case, we can say that the function $T(n)$ is of order at most n."

"Excuse me," Tetra said, raising her hand. "I've gotten a lot better with logic statements, but this is still a bit much, the N's and the C's and all... Can you give me a minute to digest all this?"

"It's not that hard," Miruka said. "Start with $T(n) \geqslant 0$. If you say that $T(n) = O(n)$, you can choose constants N and C so that the graph of function $y = T(n)$ is eventually below the function $y = Cn$."

While saying this, Miruka was sketching a simple graph.

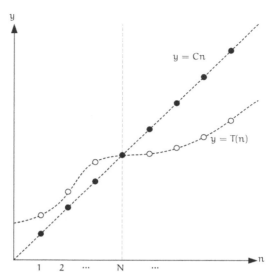

Example of $T(n)$ fulfilling $T(n) = O(n)$

"The size of $T(n)$ is being held down from above by some constant multiple of n. That's what $T(n) = O(n)$ means."

"So, like, it's making sure that $T(n)$ doesn't get too big?" Tetra said, jotting down notes.

"Right. But 'doesn't get too big' is a bad description for two reasons. One: it sounds like you're talking about exceeding some constant. In this case, it's okay for $T(n)$ to go to infinity when n does. Two: we're trying to describe things quantitatively here, so 'too big' doesn't cut it. We need a stronger statement, like 'the extent of increase in $T(n)$ is at most a constant multiple of n.' We normally shorten that to '$T(n)$ is of order at most n,' or even just '$T(n)$ is $O(n)$.'"

"What exactly does 'order' mean here?"

"It's short for 'order of growth,' the rate at which the function gets bigger."

> **Big-O notation (order of at most n)**
>
> $T(n) = O(n)$
> $\iff \exists N \in \mathbb{N}\ \exists C > 0\ \forall n \geqslant N\ \Big[|T(n)| \leqslant Cn\Big]$
> $\iff$ Function $T(n)$ is of order *at most* n

6.2.3 A Quiz

"A quiz for you," Miruka said, "using $T_L(n) = 4n+5$ as the maximum number of steps in a linear search. Tell me, why can we write this using big-O notation?"

$$4n + 5 = O(n)$$

Miruka pointed at Tetra.

"Let me see. I know something is holding down some values, but... Uh, I'm not sure."

"Hmph. You then," Miruka said, pointing at me.

"I guess I'll go back to the definition," I said. "I can set N and C both equal to 5, for example. That would mean that for all values of n that are 5 or greater, this should hold."

$$|4n + 5| \leqslant 5n$$

"So from the definition, $4n + 5 = O(n)$ holds."

"Back to the definition!" Tetra said. "I should have done that!"

"Next quiz," Miruka said. "Does this hold?"

$$n + 1000 = O(n)$$

"Oh, I can do this one!" Tetra said. "We can let $N = 1000$ and $C = 2$. Then for all values of n equal to or greater than 1000, we would get this."

$$|n + 1000| \leqslant 2n$$

"So yes, $n + 1000 = O(n)$ is true."

"Very good," Miruka said, nodding. "You can even do the same thing with the values flipped, using $N = 2$ and $C = 1000$."

"Oh, so we can use a big function like $1000n$ to hold the values down."

"Next quiz," Miruka said. "Does this equation hold?"

$$n^2 = O(n)$$

"Well, the n^2 on the left would be $1, 4, 9, 16, 25, \ldots$, so I don't think the Cn on the right can bound that."

"That's right," Miruka said. "So $n^2 = O(n)$ does *not* hold. You can't bound a quadratic function like n^2 with a multiple of n. No matter how big the C you use, once n gets big enough you'll end up with $n^2 > Cn$." Miruka looked straight at me, and she started speaking more quickly. "So we can't say that a quadratic function is of order at most n. And of course you could say the same about a cubic function, a quartic function, and so on. On the other hand, you can say that n or $n + 1000$ or $4n + 5$ *is* of at most order n."

"Categorization of functions!" I nearly shouted. "We can categorize functions according to whether they are of at most order n!"

Miruka snapped her fingers.

"Exactly. Once we've classified functions as being of at most order n, we don't have to worry about any coefficients on n, or any differences in constant terms. We can just consider them as being the same thing. That's why I can say that analysis isn't necessarily about obtaining more precision."

"Ignoring their differences..." Tetra whispered.

"Now that we've gotten a little more comfortable with the big-O notation in our toolbox," Miruka said, "let's take another look at the maximum number of execution steps in the linear search and sentinel linear search algorithms, and see if we don't notice something new."

$$T_L(n) = 4n + 5$$
$$T_S(n) = 3n + 7$$

"Well, we've said that these are both of order at most n, right?" Tetra said.

$$T_L(n) = 4n + 5 = O(n)$$
$$T_S(n) = 3n + 7 = O(n)$$

"I'm, uh, not sure what else that tells us, though."

"I guess one way to look at it would be to say that while we did reduce the number of execution steps by modifying the linear search algorithm, we only made a limited improvement. After all, the sentinel linear search algorithm is still of order at most n, so we haven't really changed its fundamental nature. Well, depending on your definition of 'fundamental,' I guess."

"Huh? So there are 'orders' other than n?" Tetra asked.

"Many," Miruka replied. "You can stick any nonnegative function you want into O( )."

6.2.4 Of Order at Most f(n)

"So let's talk about orders of some given function. In other words, we can write $T(n) = O(f(n))$."

Big-O notation (order at most $f(n)$)

$$T(n) = O(f(n))$$
$$\iff \exists N \in \mathbb{N} \ \exists C > 0 \ \forall n \geqslant N \ \left[|T(n)| \leqslant Cf(n) \right]$$
$$\iff \text{Function } T(n) \text{ is of order at most } f(n)$$

"Here's some examples."

$$\begin{aligned}
n &= O(n) & &n \text{ is at most of order } n \\
2n &= O(n) & &2n \text{ is at most of order } n \\
4n + 5 &= O(n) & &4n + 5 \text{ is at most of order } n \\
1000n &= O(n) & &1000n \text{ is at most of order } n \\
n^2 &= O(n^2) & &n^2 \text{ is at most of order } n^2 \\
2n^3 + 3n^2 &= O(n^3) & &2n^3 + 3n^2 \text{ is at most of order } n^3 \\
0.001n^{1000} &= O(n^{1000}) & &0.001n^{1000} \text{ is at most of order } n^{1000}
\end{aligned}$$

"So we can just ignore any coefficients and use the n term with the largest degree?" Tetra asked.

"In these examples, yes," Miruka said. "But remember that the definition of big-O notation uses inequalities like $|T(n)| \leqslant C f(n)$. In other words, when we're using $f(n)$ for the evaluation, we can get as big as we want."

"Meaning?"

"Meaning that using big-O notation like this is also correct."

$$n = O(n^2)$$

"Huh? You mean n is of order n^2? I thought—"

"No, this says that n is of order *at most* n^2."

"Oh, so we can overshoot like that. So I could also write something like this?"

$$n = O(n^{1000})$$

"You could," Miruka said.

"And for the $T_L(n) = 4n + 5 = O(n)$ we were talking about before, I could say this?"

$$T_L(n) = 4n + 5 = O(n^{1000})$$

"According to the definition, sure."

Lisa, who had been silently typing up to that point, suddenly looked up in surprise. Miruka glanced at her, but continued on.

"Of course, we know that we can also use $O(n)$, so using $O(n^{1000})$ instead seems like a waste of perfectly good information. Even so, there's nothing at all wrong with saying $4n + 5 = O(n^{1000})$."

"Right," Tetra said. "I guess even when I say 'at most,' somewhere inside I'm thinking 'equal to.'"

"If you want to say that $O(f(n))$ is of order *exactly* $f(n)$, use a big-Theta in place of the O."

$$\Theta(f(n))$$

"Or I could just say 'is exactly'?"

"If you want to say that $T(n)$ is of order exactly $f(n)$, we write that with big-Theta in place of the O, so $T(n) = \Theta(f(n))$. This means that $T(n)$ is both of order at most $f(n)$, written $T(n) = O(f(n))$, and of order at least $f(n)$, written with a big-Omega as $T(n) = \Omega(f(n))$.

Graphically, this means T(n) is sandwiched between two different constant multiples of f(n), at least when n > N."

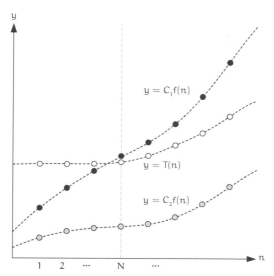

$T(n) = \Theta(f(n))$ by $T(n) = O(f(n))$ and $T(n) = \Omega(f(n))$.

Big-Θ notation (order exactly $f(n)$)

$\quad T(n) = \Theta(f(n))$
$\iff T(n) = O(f(n)) \land T(n) = \Omega(f(n))$
$\iff \exists N \in \mathbb{N} \; \exists C_1 > 0 \; \exists C_2 > 0 \; \forall n \geqslant N \; \Big[C_2 f(n) \leqslant T(n) \leqslant C_1 f(n) \Big]$
$\iff$ Function $T(n)$ is of order *exactly* $f(n)$

> **Big-Ω notation (order at least $f(n)$)**
>
> $T(n) = \Omega(f(n))$
> $\iff \exists N \in \mathbb{N} \; \exists C > 0 \; \forall n \geqslant N \; \left[T(n) \geqslant C f(n) \right]$
> $\iff$ Function $T(n)$ is of order *at least* $f(n)$

> **Big-O notation and friends**
>
> $T(n) = O(f(n))$ $T(n)$ is of order at most $f(n)$
> $T(n) = \Theta(f(n))$ $T(n)$ is of order exactly $f(n)$
> $T(n) = \Omega(f(n))$ $T(n)$ is of order at least $f(n)$

"Okay, another quiz then," Miruka said, pointing at Tetra.

> True or false: $T(n) = O(n^2)$ and $T(n) = O(3n^2)$ have the same truth value.

"False, because the coefficient 3— No, wait! True! They both have the same value!"

"Good. If $T(n) = O(n^2)$, then $T(n) = O(3n^2)$ holds, and vice versa. Similarly, $T(n) = O(n^2)$ and $T(n) = O(3n^2 + 2n + 1)$ have the same value."

"Right!"

"Okay, on to the next quiz."

> If $T(n) = O(1)$, then what kind of function is $T(n)$?

"Whoa, there's no n here, so... Oh, right, back to the definition!"

$$T(n) = O(1) \iff \exists N \in \mathbb{N} \; \exists C > 0 \; \forall n \geqslant N \; \left[|T(n)| \leqslant C \cdot 1 \right]$$

"Oh! So $T(n)$ is a constant function!"

"Wrong," Miruka said abruptly. "It works the other way, meaning that if $T(n)$ is a constant function then $T(n) = O(1)$, but the converse isn't necessarily true. Just saying that $T(n) = O(1)$ doesn't necessarily mean that $T(n)$ is a constant."

"Upper bound," Lisa mumbled, startling me.

"Right," Miruka said. "If $T(n) = O(1)$, then the function $T(n)$ will never exceed some specific value, no matter how large n gets. In other words, it has an upper bound. It doesn't matter how or how much it changes, so long as it never exceeds that bound."

"So you mean that $T(n)$ converges to some value as n goes to infinity?" Tetra asked.

"No," Miruka said. "$T(n)$ might go up and down or whatever, so long as it stays below the upper limit. $T(n) = (-1)^n$ would be a good example. So saying that $T(n) = O(1)$ doesn't necessarily mean that $T(n)$ will converge."

Listening to this back and forth between Miruka and Tetra, I was filled with an indescribable happiness. From the perspective of a mathematician or a computer scientist, our discussion of big-O notation, the equations and logic we used and the mathematical arguments we made, would probably be unworthy of attention. But for me, this was the best thing in the world.

6.2.5 $\log n$

Tetra raised her hand. "So does big-O notation always come in one of these forms?"

$$O(1), \ O(n), \ O(n^2), \ O(n^3), \ O(n^4), \ \ldots$$

"There's plenty more," Miruka said. "There's even infinitely many orders between 1 and n. In fact, $\log n$ is a very important one."

$$O(\log n)$$

"An order of $\log n$ is smaller than an order of n. Asymptotically speaking, if the maximum number of execution steps is on the order of $\log n$, then it's a very good algorithm." Miruka started twirling a finger as she continued. "Similarly, between n and n^2 there's also an order of $n \log n$."

$$O(n \log n)$$

"Be careful, you should read that as $n \times \log(n)$, not $n \times \log \times n$," I said.

"Sure, I know. The 'log' here stands for 'logarithm,' right?"

"The logarithm function, yes," Miruka said. "A function that returns the logarithm of a given n. When you're talking about logarithms in general you normally have to specify a base, like $\log_2 n$ or $\log_{10} n$ or $\log_e n$. There's no need when talking about big-O notation, though, because all logarithmic functions are just constant multiples of each other, even when their bases change."

"You convert bases like this, right?" I said.

$$\begin{aligned}\log_A x &= \log_A B^{\log_B x} && \text{because } x = B^{\log_B x} \\ &= (\log_B x) \cdot (\log_A B) && \text{because } \log_A B^\alpha = \alpha \cdot \log_A B \\ \log_A x &= \underbrace{(\log_A B)}_{\text{constant}} \cdot (\log_B x) && \log_A x \text{ is a constant multiple of } \log_B x\end{aligned}$$

"Yep," Miruka said with a nod. "So we don't have to worry about what base we're using when writing big-O notation."

$$\begin{aligned} T(n) &= O(\log_2 n) \\ \Longleftrightarrow T(n) &= O(\log_{10} n) \\ \Longleftrightarrow T(n) &= O(\log_e n) \end{aligned}$$

"I see," I said.

"Also remember that the log function is the inverse of the exponential function. Exponential functions increase very rapidly. Conversely, logarithmic functions rise very slowly. For example, using a base of 2, 2^n increases to $2^1 = 2, 2^2 = 4, 2^3 = 8$, and so on, but $\log_2 n$ increases little by little. Functions like $\log_2 n$ that increase only very slowly, in other words functions with a small order of growth, mean that even when the size of the input becomes enormous, the algorithm can still complete with only a relatively small increase in the number of execution steps. In other words, it is an asymptotically fast algorithm."

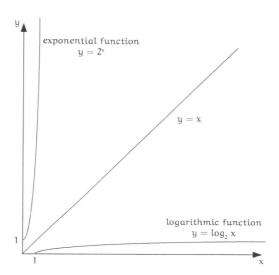

"Are there really algorithms that can complete within $O(\log n)$ steps?" Tetra asked.

"Of course. The binary search algorithm can find its target with order exactly $\log n$, for example."

"Hang on," I said. "There's something strange about that. If you search an array for something it doesn't contain, you have to look at all n entries to be sure it doesn't, right? How could you do that with order $\log n$? Doesn't a full search mean you can never do better than order n?"

"Not under the right conditions," Miruka said. "Tetra, I'll bet that card Mr. Muraki gave you this morning is nudging us in the right direction."

"Oh, right! I'd almost forgot!" Tetra said, digging through her bag. She pulled out a card and held it up for us to see.

6.3 SEARCH

6.3.1 Binary Search

> **The binary search algorithm (input and output)**
>
> Input
> - Array $A = \langle A[1], A[2], A[3], \ldots, A[n] \rangle$,
>   where $A[1] \leqslant A[2] \leqslant A[3] \leqslant \cdots \leqslant A[n]$.
> - Array size n
> - Search target v
>
> Output
> When an element equal to v is found in A,
>   output $\langle \text{FOUND} \rangle$
> When no element equal to v is found in A,
>   output $\langle \text{NOT FOUND} \rangle$

"So this is the input and output for a binary search?" Tetra said. "We're looking for a value v in an array A, so I guess this is just like when we did a linear search."

"Not quite," I said, pointing at the card. "These conditions are new."

$$A[1] \leqslant A[2] \leqslant A[3] \leqslant \cdots \leqslant A[n]$$

"Oh, right! But...what do these mean?"

"Looks like the values in the array are sorted into ascending order. The smallest value comes first, and every next element is equal to or larger than the one before it."

"Aha! And I'll bet this is key to lowering the order of our search! Oh, and I got one other card, too. Here it is."

> **The binary search algorithm (procedure)**
>
> $C1$:   procedure BINARY-SEARCH(A, n, v)
> $C2$: $a \leftarrow 1$
> $C3$: $b \leftarrow n$
> $C4$:   while $a \leqslant b$ do
> $C5$: $k \leftarrow \lfloor \frac{a+b}{2} \rfloor$
> $C6$:   if $A[k] = v$ then
> $C7$:   return $\langle$ FOUND $\rangle$
> $C8$:   else-if $A[k] < v$ then
> $C9$: $a \leftarrow k + 1$
> $C10$:   else
> $C11$: $b \leftarrow k - 1$
> $C12$:   end-if
> $C13$:   end-while
> $C14$:   return $\langle$ NOT FOUND $\rangle$
> $C15$: end-procedure

Tetra and I took some time to read through this new card. Lisa glanced at it, then turned back to her computer. Miruka was staring out the window, tapping her jaw.

"This is... hard," Tetra said. "I think I need to do a line-by-line walkthrough to be sure I understand what's going on."

"Do you understand line $C5$?" Miruka said, still at the window.

"I think so. It looks like we're storing the average of a and b in k."

"You're skipping over an important part of that line—the floor function, indicated by the brackets, which says to discard anything after the resulting decimal point. $\frac{a+b}{2}$ is the midpoint between values a and b, which could be anywhere. If $a + b$ is an odd number then dividing by two would not give you an integer, so you couldn't store it in k."

"Gotcha," Tetra said, writing in her notebook. "So these $\lfloor \ \rfloor$ brackets are called 'floor,' and they mean to round down."

"Correct. You can also say that $\lfloor x \rfloor$ is the largest integer that is not greater than x. If x is an integer then $\lfloor x \rfloor$ is just x. Other examples would be $\lfloor 3 \rfloor = 3, \lfloor 2.5 \rfloor = 2, \lfloor -2.5 \rfloor = -3, \lfloor \pi \rfloor = 3$."

"Okay," Tetra said, turning to a new page. "Now I guess I'm ready to create a test case and do my walkthrough!"

6.3.2 An Example

We gave Tetra some time to trace through the algorithm, and after a time she looked up from her notebook.

"Okay," she said, "I think I've got it, but let's run through it together. So here's my setup. We're going to search array A for the value 77."

$$A = \langle 26, 31, 41, 53, 77, 89, 93, 97 \rangle, \; n = 8, \; v = 77$$

$C1$:	procedure BINARY-SEARCH(A, n, v)	$\boxed{1}$
$C2$:	$a \leftarrow 1$	$\boxed{2}$
$C3$:	$b \leftarrow n$	$\boxed{3}$
$C4$:	while $a \leqslant b$ do	$\boxed{4}\;\boxed{11}\;\boxed{18}$
$C5$:	$k \leftarrow \lfloor \frac{a+b}{2} \rfloor$	$\boxed{5}\;\boxed{12}\;\boxed{19}$
$C6$:	if $A[k] = v$ then	$\boxed{6}\;\boxed{13}\;\boxed{20}$
$C7$:	return $\langle$ FOUND $\rangle$	$\boxed{21}$
$C8$:	else-if $A[k] < v$ then	$\boxed{7}\;\boxed{14}$
$C9$:	$a \leftarrow k + 1$	$\boxed{8}$
$C10$:	else	
$C11$:	$b \leftarrow k - 1$	$\boxed{15}$
$C12$:	end-if	$\boxed{9}\;\boxed{16}$
$C13$:	end-while	$\boxed{10}\;\boxed{17}$
$C14$:	return $\langle$ NOT FOUND $\rangle$	
$C15$:	end-procedure	$\boxed{22}$

Binary search walkthrough
(Input: $A = \langle 26, 31, 41, 53, 77, 89, 93, 97 \rangle, n = 8, v = 77$)

"So we start with $\boxed{5}$, where $a = 1$, $b = 8$, and $k = \lfloor \frac{1+8}{2} \rfloor = \lfloor 4.5 \rfloor = 4$. Then in $\boxed{6}$ and $\boxed{7}$ we compare $A[k] = A[4] = 53$ and $v = 77$."

$$
\begin{array}{cccc}
a & & k & b \\
\Downarrow & & \Downarrow & \Downarrow
\end{array}
$$

1	2	3	4	5	6	7	8
26	31	41	53	77	89	93	97

"Things get interesting at $\boxed{7}$. Here $53 < 77$, so $A[k] < v$ holds. This meets the requirement of $\boxed{7}$, so at $\boxed{8}$ we perform the substitution $a \leftarrow k + 1$. What we're doing here is making a larger, which greatly narrows the range over which we have to search! If $A[k] < v$, then if v is here somewhere it must be to the right of $A[k]$, so we don't have to look anywhere to its left.

"Okay, on to $\boxed{12}$, where we have $a = 5$, $b = 8$, and $k = \lfloor \frac{5+8}{2} \rfloor = \lfloor 6.5 \rfloor = 6$. Then at $\boxed{13}$ and $\boxed{14}$ we compare $A[k] = A[6] = 89$ and $v = 77$."

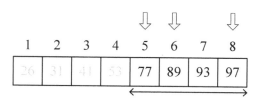

"Since $77 < 89$, $v < A[k]$ holds, which means that the conditions in $\boxed{13}$ and $\boxed{14}$ are not met. So we go to $\boxed{15}$, where we perform the substitution $b \leftarrow k - 1$. Again, we've narrowed the range we have to search over, but this time we're going the other way, making b smaller.

"Finally we go to $\boxed{19}$, where we have $a = 5$, $b = 5$, and $k = \lfloor \frac{5+5}{2} \rfloor = \lfloor 5 \rfloor = 5$. Then at $\boxed{20}$ we compare $A[k] = A[5] = 77$ and $v = 77$, and see we've found our target v! Yay!"

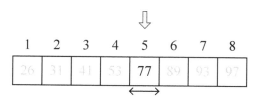

"There sure were a whole lotta steps in all that," I said. "Was that really faster?"

"Like Tetra showed us, it's all about the a-or-greater-b-or-less search range that we're using," Miruka said. "We can look at the $A[k]$ and v comparisons to see how the speed is increasing. Every time we perform a comparison the search range gets cut in half. In other words, every increase in the number of comparisons performed means we can accommodate a doubly large array size for our search. Conversely, if we're searching an array of size n, we can limit the number of comparisons to just $\log_2 n$. Binary search is a truly wonderful algorithm."

"Interesting," I said. "But how do we show that it really is of order $O(\log n)$?"

6.3.3 Analysis

Problem 6-1 (execution steps in a binary search)

Show that the number of execution steps in the BINARY-SEARCH algorithm is $O(\log n)$.

"Think you can be convinced without equations?" Miruka said to me, slyly grinning. "Let's analyze our binary search."

"Already done," Lisa said, turning her display to us.

	Steps	Binary search
$C1$:	1	procedure BINARY-SEARCH(A, n, v)
$C2$:	1	$a \leftarrow 1$
$C3$:	1	$b \leftarrow n$
$C4$:	$M+1$	while $a \leqslant b$ do
$C5$:	$M+S$	$\quad k \leftarrow \lfloor \frac{a+b}{2} \rfloor$
$C6$:	$M+S$	$\quad$ if $A[k] = v$ then
$C7$:	S	$\quad\quad$ return $\langle$ FOUND $\rangle$
$C8$:	M	$\quad$ else-if $A[k] < v$ then
$C9$:	X	$\quad\quad a \leftarrow k+1$
$C10$:	0	$\quad$ else
$C11$:	Y	$\quad\quad b \leftarrow k-1$
$C12$:	M	$\quad$ end-if
$C13$:	M	end-while
$C14$:	$1-S$	return $\langle$ NOT FOUND $\rangle$
$C15$:	1	end-procedure

Analysis of procedure BINARY-SEARCH

Number of execution steps in binary search
$$= C1 + C2 + C3 + C4 + C5 + C6 + C7 + C8 + C9 + C10$$
$$+ C11 + C12 + C13 + C14 + C15$$
$$= 1 + 1 + 1 + (M+1) + (M+S) + (M+S) + S + M + X + 0$$
$$+ Y + M + M + (1-S) + 1$$
$$= 6M + X + Y + 2S + 6$$

"Well done, Lisa," Miruka said. "But we know that $X + Y = M$, so we can go one step further."

Number of execution steps in a binary search $=$ $7M + 2S + 6$

"There you go. The number of execution steps in a binary search is $7M + 2S + 6$. The S is an indicator that takes a value of 1 when we've found what we're looking for, so it will be either 0 or 1. The biggest influence on the result is M, which equals the number of comparisons we performed at step $C8$. So all we need to be able to say is that the maximum number of comparisons M at step *C8* is $O(\log n)$."

"M is $O(\log n)$ ..." I said. "Huh. But hang on, M is going to change with the input size n, so wouldn't it be better to express that as a function, like $M(n)$?"

Miruka nodded. "You could do that, sure."

"So letting $M(n)$ be the maximum number of comparisons at step *C8*, the expression we want to prove is this."

$$M(n) = O(\log n)$$

Tetra raised her hand. "I think I understand the $M(n) = O(\log n)$ here, but what kind of function is $M(n)$?"

"You tell us." Miruka said with a wink.

"Who, me? Uh, okay. So I guess I'll create another example, using specific values for $A[1], A[2], A[3]\ldots$"

"Hold up, Tetra," I said. "If we're going to do a good walkthrough, first we need an appropriate test case for considering the *maximum* number of comparisons at step *C8*. If you just make one up, you might just happen to find one that requires fewer than normal comparisons."

"No problem!" Tetra said. "All we have to do is search for a value that isn't in the array. That should guarantee the maximum number of comparisons, right?"

"No, not necessarily. Remember what step *C5* said? We're rounding down when we do the substitution $k \leftarrow \lfloor \frac{a+b}{2} \rfloor$, so we should specifically look for a number that's *bigger* than any number in the array. That will always make us search to the end of the right half of the array."

"Oh, I get it! Because the right half is all bigger numbers than the left half, right? Okay, I'll make an example where we're looking for a number that's bigger than anything we could find in the array."

"Yeah, I'll do the same."

We both worked for a while, when Tetra startled me with a shout.

"Look! Look! I found a cool pattern! For example, say that $n = 16$, and we're looking for a value v bigger than anything in the array. The binary search looks at array elements in this order."

Tetra and I worked to create a table of $M(n)$ values for small values of n. We noticed the pattern quicker than I expected, so creating the table didn't take long.

n	Position of compared element (■)	$M(n)$ (No. of ■s)
1	■	1
2	■■	2
3	□■■	2
4	□■■■	3
5	□□■■■	3
6	□□■□■■	3
7	□□□■□■■	3
8	□□□■□■■■	4
9	□□□□■□■■■	4
10	□□□□■□□■■■	4
11	□□□□□■□□■■■	4
12	□□□□□■□□■□■■	4
13	□□□□□□■□□■□■■	4
14	□□□□□□■□□□■□■■	4
15	□□□□□□□■□□□■□■■	4
16	□□□□□□□■□□□■□■■■	5
⋮	⋮	⋮

Relation between input size n and maximum number of comparisons at step C8

"Nice. The regularity is very clear," Miruka said.

"It is," I agreed. "We can show the relation between n and $M(n)$ like this."

$$2^{M(n)-1} \leqslant n$$

"Really?" Tetra said. She looked back and forth between the inequality and the table. "Oh, yeah. I see it now. So when do I get to the point where I can just whip out math like that?"

"If the base in this inequality is 2, we can approach the target equation," I said.

$$2^{M(n)-1} \leqslant n \qquad \text{prediction}$$
$$\log_2 2^{M(n)-1} \leqslant \log_2 n \qquad \text{take log 2 of both sides}$$
$$M(n) - 1 \leqslant \log_2 n \qquad \text{from the definition of logarithms}$$
$$M(n) \leqslant 1 + \log_2 n \qquad \text{move 1 to the right}$$
$$M(n) = O(\log_2 n) \qquad \text{ignore constant terms}$$
$$M(n) = O(\log n) \qquad \text{ignore logarithmic base}$$

"That's it?" Tetra said. "That's the proof?"

"We've still got to deal with the prediction part, the $2^{M(n)-1} \leqslant n$. In other words, $M(n) \leqslant 1 + \log_2 n$. It's nearly trivial, but I guess it's provable by mathematical induction."

I paused to consider how to tackle this, then continued.

"We want to show that $M(n) \leqslant 1 + \log_2 n$ for $n = 1, 2, 3, \ldots$ We know this holds for $n = 1$, because $M(1) = 1$ on the left, and $1 + \log_2 1 = 1$ on the right. Next we assume that it holds for $n = 1, 2, 3, \ldots, j$ and show that it also holds for $n = j + 1$. To make it easier to understand, let's do separate cases for when $n = j + 1$ is even and odd."

When $n = j + 1$ is even...

"When we perform the comparison at the black square and move on to search the $\frac{j+1}{2}$ squares on the right, we get an inequality like this."

$$M(j+1) \leqslant 1 + M\left(\frac{j+1}{2}\right) \qquad \text{comparison and right-side search}$$
$$\leqslant 1 + \left(1 + \log_2 \frac{j+1}{2}\right) \qquad \text{inductive step}$$
$$= 2 + \log_2(j+1) - \log_2 2 \qquad \text{from properties of logarithms}$$
$$= 1 + \log_2(j+1)$$

"So now we know that this holds."

$$M(j+1) \leqslant 1 + \log_2(j+1)$$

When $n = j+1$ is odd...

"This time, when we make the comparison at the black square and search the $\frac{j}{2}$ squares on the right, we get an inequality like this."

$$\begin{aligned}
M(j+1) &\leqslant 1 + M\left(\frac{j}{2}\right) && \text{comparison and right-side search} \\
&\leqslant 1 + \left(1 + \log_2 \frac{j}{2}\right) && \text{inductive step} \\
&= 2 + \log_2 j - \log_2 2 && \text{from properties of logarithms} \\
&= 1 + \log_2 j \\
&\leqslant 1 + \log_2(j+1)
\end{aligned}$$

"So this must hold, regardless of whether $n = j+1$ is even or odd."

$$M(j+1) \leqslant 1 + \log_2(j+1)$$

"And from the principle of mathematical induction, we get

$$M(n) \leqslant 1 + \log_2 n$$

for all values of n."

Answer 6-1 (execution steps in a binary search)

The BINARY-SEARCH procedure in the binary search algorithm completes execution in $O(\log n)$ steps.

6.3.4 A Promise of Sorts

"Rainbow," Lisa said, looking out the window.

"Ooh! It is!" Tetra said, running to look outside. Sure enough, there was a rainbow there, though so faint it seemed it might fade away at any time.

"We did have a spring shower a little while ago," Miruka said.

"A rainbow is a symbol of a promise. Very fitting, since we've just demonstrated a promise that the number of execution steps in a binary search will be at most of order $\log n$!"

"Huh," I said, taking a second look at the fading rainbow and thinking again about promises. Promises made, promises kept, promises broken... *"We'll do more math tomorrow."*

"Oh, I just thought of something!" Tetra said. "Now we know that a binary search is guaranteed to finish in $O(\log n)$ time, but that assumes that the array is sorted, right? So we can do fast searches if we have a sorted array, but—"

"—but we can't really say we have a fast search algorithm if it's going to take a long time to sort the array," Miruka said.

"Right!"

"Of course, if we know we will be repeatedly searching an array it makes sense to sort it beforehand. There are lots of ways to do sorts. I don't suppose you have a card for one?"

Tetra flipped over the next card in her stack as if performing a magic trick.

"Something about... bubbles?"

6.4 SORTING

6.4.1 Bubble Sort

The bubble sort algorithm (input and output)

Input

- Array $A = \langle A[1], A[2], A[3], \ldots, A[n] \rangle$
- Array size n

Output

An array with the elements of the input array sorted in ascending order

$$A[1] \leqslant A[2] \leqslant A[3] \leqslant \cdots \leqslant A[n]$$

"So sorting an array means rearranging its elements in some order, right?" Tetra asked.

"Right," Miruka replied. "Look at the input and output on your card. In this case we specifically want the numbers to be sorted in ascending order, from smallest to largest."

$$A[1] \leqslant A[2] \leqslant A[3] \leqslant \cdots \leqslant A[n]$$

"To fulfill these conditions, we need to rearrange the elements of the input array. Sort them, in other words."

"Got it. Okay, let's read through the procedure."

The bubble sort algorithm (procedure)

B1:	procedure BUBBLE-SORT(A, n)
B2:	$m \leftarrow n$
B3:	while $m > 1$ do
B4:	$k \leftarrow 1$
B5:	while $k < m$ do
B6:	if $A[k] > A[k+1]$ then
B7:	$A[k] \leftrightarrow A[k+1]$
B8:	end-if
B9:	$k \leftarrow k + 1$
B10:	end-while
B11:	$m \leftarrow m - 1$
B12:	end-while
B13:	return A
B14:	end-procedure

"Hmm, what's going on in this line $B7$, $A[k] \leftrightarrow A[k+1]$?" Tetra asked, flipping back through her notes. "Oh, wait, here it is. We swap the values of two variables, $A[k]$ and $A[k+1]$ in this case, right?"

We fell silent for a time as we did our own mental walkthroughs, trying to figure out just what this algorithm was doing. Only by becoming computers ourselves and stepping through line-by-line could we really reach a deep understanding. At least, that's how it was supposed to work.

6.4.2 An Example

"I think I get it!" Tetra said. "We look for adjacent pairs of elements in the wrong order and flip them around, and we're repeating that $n-1$ times. If we perform BUBBLE-SORT using $A = \langle 53, 89, 41, 31, 26 \rangle$ and $n = 5$ as a test case, it goes like this."

Tetra opened her notebook and showed us her walkthrough.

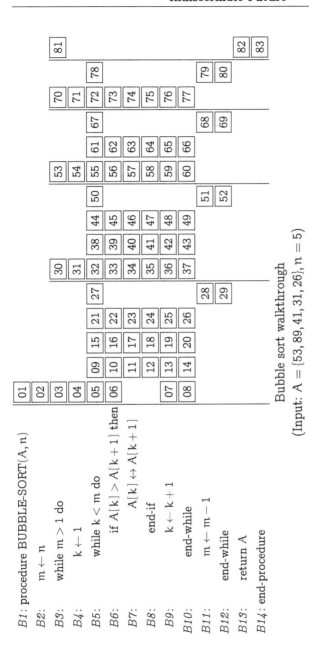

Bubble sort walkthrough
(Input: $A = \{53, 89, 41, 31, 26\}$, $n = 5$)

"Why's it called a bubble sort, anyway?" Tetra asked.

"Because the first values in the array float up, like bubbles in water," Miruka said.

	m=5				m=4			m=3		m=2
	k=1	k=2	k=3	k=4	k=1	k=2	k=3	k=1	k=2	k=1
A[1]	<u>53</u> → 53	53	53	<u>53</u>	41	41	<u>41</u>	31	31	26
A[2]	<u>89</u> → <u>89</u>	41	41	<u>41</u>	53	31	<u>31</u>	<u>41</u>	26	31
A[3]	41	<u>41</u>	<u>89</u>	31	31	<u>31</u>	<u>53</u>	26	<u>26</u>	41	41
A[4]	31	31	<u>31</u>	<u>89</u>	26	26	<u>26</u>	53	53	53	53
A[5]	26	26	26	<u>26</u>	89	89	89	89	89	89

6.4.3 Analysis

"I wonder if there's some way to analyze the bubble sort." I said.

"Actually, I think I've got an analysis," Tetra murmured. "Part of one, at least. Here's as far as I got."

	Times executed	Bubble sort
$B1$:	1	procedure BUBBLE-SORT(A, n)
$B2$:	1	$m \leftarrow n$
$B3$:	n	while $m > 1$ do
$B4$:	$n-1$	$\quad k \leftarrow 1$
$B5$:		$\quad$ while $k < m$ do
$B6$:		$\quad\quad$ if $A[k] > A[k+1]$ then
$B7$:		$\quad\quad\quad A[k] \leftrightarrow A[k+1]$
$B8$:		$\quad\quad$ end-if
$B9$:		$\quad\quad k \leftarrow k+1$
$B10$:		$\quad$ end-while
$B11$:	$n-1$	$\quad m \leftarrow m-1$
$B12$:	$n-1$	end-while
$B13$:	1	return A
$B14$:	1	end-procedure

(Partial) analysis of procedure BUBBLE-SORT

"Huh? Are you sure line $B3$ here only gets executed n times?" I asked.

"I think so," Tetra said, pointing at the card to explain. "Just before we get to line $B3$ the computer has executed either line $B2$ or $B12$, right? So the number of executions of line $B3$ should be the sum of the counts for $B2$ and $B12$. When we get to line $B3$ from $B2$ that's one execution, and we go back there from line $B12$ $n-1$ times. Add them together, and that's n executions!"

"Interesting," I said.

"So that's how many times we enter line $B3$. After that, the computer goes on to either line $B4$ or line $B13$. It should go to line $B4$ $n-1$ times and to line $B13$ just once. So again, add them up and it's n times, which is the number of times we leave line $B4$. The number of times we enter that line and the number of times we leave it agree, so I'm pretty sure the number of executions of line $B3$ is n."

"Well done," I said, impressed by Tetra's clear explanation. "It makes perfect sense that the number of times you enter a line should be the same number of times you leave it, but even so it's an interesting property."

"Kirchoff's first law," Lisa muttered.

"It has a name?" I said, turning to Lisa, but got no response. I turned back to Tetra. "Shouldn't the count in line $B5$ be $n-1$?"

"No, I don't think so. Everything from line $B5$ to line $B10$ is in in a 'while' statement that's inside another 'while' statement, so they're going to get repeated a whole bunch."

"Ah, indeed they will," I said. "I wonder if we can get a bit more quantitative than 'a whole bunch,' though."

Problem 6-2 (Analysis of the maximum number of execution steps in a bubble sort)

Use big-O notation to represent the maximum number of execution steps when performing the BUBBLE-SORT procedure over an array of size n.

Looking over Tetra's notes, I said, "This line $B5$ here, first it's going to get executed five times, then four, then three, then two times, right?"

"Sure looks that way," Tetra said. "The value of m here goes down by one each time through."

"Right. So that means line $B5$ will be executed $n + (n-1) + (n-2) + \cdots + 3 + 2$ times."

"Oh, yeah! So we just have to add up the numbers from 2 to n."

"Which is equal to $\frac{n(n+1)}{2} - 1$."

$$\begin{aligned}\text{Executions of line } &B5 \\ &= n + (n-1) + (n-2) + \cdots + 2 \\ &= \frac{n(n+1)}{2} - 1 \\ &= \frac{1}{2}n^2 + \frac{1}{2}n - 1\end{aligned}$$

"And now that we know the number of times line $B5$ executes, we know the number for lines $B6$ through $B10$!"

"We do. In your test case there won't be any swaps at line $B7$, but we should assume there would be if we're going to think about the maximum number of executions that might be possible. Okay, so let's let B be the maximum number of executions of line $B5$."

$$B = \frac{1}{2}n^2 + \frac{1}{2}n - 1$$

"This makes things easy! We can say the number of steps is $B - (n-1) = B - n + 1$ for lines $B6$ through $B10$."

	Times executed	Bubble sort
B1:	1	procedure BUBBLE-SORT(A, n)
B2:	1	m ← n
B3:	n	while m > 1 do
B4:	n − 1	k ← 1
B5:	B	while k < m do
B6:	B − n + 1	if A[k] > A[k + 1] then
B7:	B − n + 1	A[k] ↔ A[k + 1]
B8:	B − n + 1	end-if
B9:	B − n + 1	k ← k + 1
B10:	B − n + 1	end-while
B11:	n − 1	m ← m − 1
B12:	n − 1	end-while
B13:	1	return A
B14:	1	end-procedure

Analysis of procedure BUBBLE-SORT (maximum execution steps)

Maximum execution steps in a bubble sort

$= B1 + B2 + B3 + B4 + B5 + B6 + B7$
$\quad + B8 + B9 + B10 + B11 + B12 + B13 + B14$
$= 1 + 1 + n + (n-1) + B + (B-n+1) + (B-n+1)$
$\quad + (B-n+1) + (B-n+1) + (B-n+1) + (n-1) + (n-1) + 1 + 1$
$= 6B - n + 6$
$= 6\left(\frac{1}{2}n^2 + \frac{1}{2}n - 1\right) - n + 6 \quad$ (because $B = \frac{1}{2}n^2 + \frac{1}{2}n - 1$)
$= 3n^2 + 2n$

"And the $3n^2 + 2n$ means it's of order at most n^2," I said.

Maximum execution steps in a bubble sort $= O(n^2)$

"Looks like we got it!" Tetra said.

> **Answer 6-2 (Analysis of the maximum number of execution steps in a bubble sort)**
>
> The maximum number of execution steps when performing the BUBBLE-SORT procedure over an array of size n is
> $$O(n^2)$$

6.4.4 Big-O hierarchy

"One other thing," Miruka said. "When you use big-O notation, you can't mix and match equalities. For example, $4n + 5 = O(n)$ and $3n + 7 = O(n)$, but that doesn't mean $4n + 5 = 3n + 7$."

"Makes sense," I said, nodding.

"Which I guess means the equals sign in big-O notation means something different from what we're used to," Tetra said.

Miruka nodded. "Because $O(f(n))$ is a set of functions. Specifically, the set of functions fulfilling the condition $|T(n)| \leqslant C\,f(n)$. A formal definition would be this."

$$\{g(n) \mid \exists N \in \mathbb{N}\ \exists C > 0\ \forall n \geqslant N\ |g(n)| \leqslant C\,f(n)\}$$

"If it's a set, we can say that $T(n) = O(f(n))$ means $T(n) \in O(f(n))$?"

"Exactly. Which should explain why the equals sign in $T(n) = O(f(n))$ doesn't allow for left–right exchanges."

"Sure does," Tetra said.

"Thinking of big-O notation as a description of sets also establishes a hierarchy on the set of functions, based on inclusion of sets. $O(1)$ is contained within $O(\log n)$, which is contained within $O(n)$, and so on."

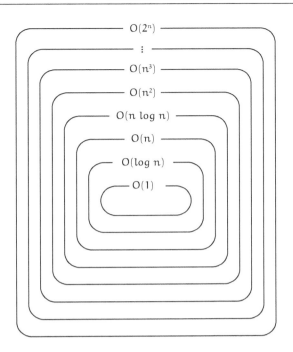

"With big-O notation, we've obtained a good way of describing the speed of an algorithm. Not to confuse speed in this sense with how fast a program will run on an actual computer. We're ignoring coefficients and constant terms, don't forget. Even so, using big-O notation as a kind of shorthand lets us describe the behavior of an algorithm when the size of its input is extremely large, in other words its asymptotic behavior. If you write out an algorithm in pseudocode one way maybe it would complete in $4n + 5$ steps, and if you wrote it another way maybe it would take $3n + 1$ steps. The order hasn't changed, though—it's n in both cases. And in both cases we can say the order is $O(n)$ to describe the asymptotic behavior of our algorithm."

6.5 Dynamic Perspectives, Static Perspectives

6.5.1 Number of Needed Comparisons

"I'm not sure how to explain it," Tetra said, "but I feel like I'm using different parts of my brain when I'm tracing through an algorithm and when I'm trying to describe the number of execution steps as a mathematical expression."

"Actually, I know exactly what you mean," I said.

"That's because it's easier to analyze something that's static than something that's dynamic," Miruka said.

"Dynamic, in what sense?" Tetra asked.

"Dynamic in the sense of a walkthrough, where you need time and a sequence to trace. Something that's static, on the other hand, allows you to see its entire structure at a glance."

"Which is why I'm more comfortable when I can write something down as equations," I said. "It's all there, right in front of me."

"Sometimes we can convert the dynamic into something static, and perform our analysis on that." Miruka quickly said. "We can cast aside time and order to produce something that's easier to handle. It's a powerful technique... when it works."

"How do you do it, though?" Tetra asked.

"Hmph. Here's a classic problem."

Problem 6-3 (Maximum number of comparisons in a comparative sort)

When performing a comparative sort on an array of size n, will the maximum number of comparisons be of order at least $n \log n$? In other words, letting $T_{\max}(n)$ be the maximum number of comparisons, can we say that

$$T_{\max}(n) = \Omega(n \log n)$$

for any comparative sort algorithm? Assume that the array contains no duplicated elements.

"Um, what's a comparative sort?" Tetra asked.

"Any sorting method where you compare the magnitudes of two elements. A bubble sort is one example."

"Gotcha. Okay, then let me count comparisons! Just give me a few minutes and I'm sure—"

"Whoa whoa whoa whoa, hold up there Tetra," I said.

"Four 'whoas', that's a prime! Wait, no it isn't ... Anyway, what's wrong?"

"How exactly do you plan to do your counting?"

"Like we did for the bubble sort, I thought."

"Not good enough. We have to show that the maximum number of comparisons will be $\Omega(n \log n)$ for *any* kind of comparative sort, not just a bubble sort."

"Oops."

"If we can show that, it would mean that no matter how brilliant a programmer you are, no matter how clever your sorting algorithm, you can't sort an array at an order of less than $n \log n$."

"Wow, that's saying quite a lot."

"By limiting the discussion to comparative sorts," Miruka said, "we're saying that the basic operation has to be comparing two elements to see which is smaller. The processing will vary with each judgment, which makes it difficult to consider every possible case—that would be an attempt to capture it dynamically. Instead, we should capture it statically, and to do that we use comparison trees."

6.5.2 Comparison Trees

"What's a comparison tree?" Tetra asked.

"Say we have three elements $A[1], A[2], A[3]$ and we are using some comparative algorithm to sort them. We can create a comparison tree like this to show the size relations between them."

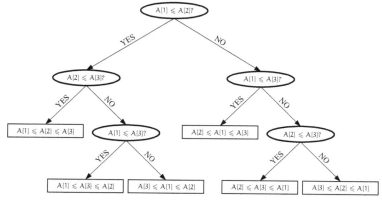

Example comparison tree

"The internal nodes in the comparison tree, the ovals, show comparisons of two elements in the array. The topmost internal node here is called the root. We start at the root and follow the branches in order, according to which elements we compare. In a comparison tree, we focus only on which two elements are compared to determine their order—we're ignoring exactly how we move elements within the array. So a comparison tree is a way of statically showing between-element comparisons as a tree structure."

"What about these rectangles?" Tetra asked.

"Those are called leaves, because they always come at the end of a branch. In a comparison tree, they show the order of all the elements, which was determined as we travelled there from the root. Every possible arrangement of the array elements should be in one of the leaves, because we need to be able to sort any array that's passed to the algorithm. Looking at it another way, if there are arrangements of array elements that can't be found among the tree's leaves, that would imply the existence of arrays that the algorithm cannot sort."

Miruka paused to look at each of us.

"The number of inner nodes passed when moving from root to leaf is equivalent to the number of comparisons the algorithm performs. That value is called the 'height' of the comparison tree, or sometimes its 'depth.' This height is the maximum number of inner nodes, which in this comparative sort algorithm equals the maximum number of comparisons. The height of this particular comparative tree is 3. And having created this tree, we have transformed the dynamic behavior of the comparative sort into the static nature of a comparative tree."

"Nice," I said. "And this lets us forget about all the possible cases that arise during sorting. Instead, we just have to look at the structure of the comparison tree."

"Now that we understand comparison trees," Miruka said, "the rest is simple. Every inner node branches into exactly two possibilities. So what does that say about the number of leaves in a comparison tree of height h? We have 2 branchings occurring at most h times, so the number of leaves will be at most 2^h. Also, there are $n!$ ways to arrange an array with n elements. Since we've said that every possible arrangement of the array must be in one of the tree's leaves, this inequality should hold."

$$2^h \geqslant n!$$

"Let's take the log base 2 of both sides. The logarithmic function $y = \log_2 x$ is a monotonically increasing function, so the direction of the inequality sign doesn't change, and we get this."

$$\log_2 2^h \geqslant \log_2 n!$$

"In other words..."

$$h \geqslant \log_2 n!$$

"The height h of the comparison tree is equal to the maximum number of comparisons, so to prove $h = \Omega(n \log n)$, we just have to show that $\log_2 n! = \Omega(n \log n)$."

6.5.3 Evaluating $\log n!$

"Interesting," I said. "To evaluate $\log_2 n!$, we just need to estimate $n!$, right? We can use, uh, what did you call it? Stirling's approximation? How did that go..."

"We don't need Stirling's approximation to do what we want to do," Miruka said, "namely to prove that $\log_2 n! = \Omega(n \log n)$. A much rougher approximation will suffice."

"Question!" Tetra said, raising her hand. "What exactly do you two mean by 'evaluate' here?"

"Here we mean to estimate how big something is," I said.

"How about we give Tetra a specific example," Miruka said. "For example, let's estimate $\log 6!$."

$$6! = 6 \cdot 5 \cdot 4 \cdot 3 \cdot 2 \cdot 1$$

"Okay, sounds good!" Tetra said.

"We want to construct a number that's smaller than this one, to bound $6!$ from below. For example, 3 is near the middle of these numbers, $6, 5, 4, 3, 2, 1$, so we can compare $6!$ with 3 cubed."

$$6! = 6 \cdot 5 \cdot 4 \cdot 3 \cdot 2 \cdot 1 \geqslant 6 \cdot 5 \cdot 4 \geqslant 3 \cdot 3 \cdot 3 = 3^3$$

"Yeah, sure. This is just an example, right?"

"It is. An example that gives us this."

$$6! \geqslant 3^3$$

"Sure, I see that."

"If we take the log of both sides, we get this."

$6! \geqslant 3^3$

$\log 6! \geqslant \log 3^3$ take the log

$\log 6! \geqslant 3 \log 3$ because $\log 3^3 = 3 \log 3$

"Knowing that I'm going to want to generalize this, I'm going to rewrite the 3s here as $\frac{6}{2}$."

$$\log 6! \geqslant \frac{6}{2} \log \frac{6}{2}$$

"Now we can generalize everything up to now for base 2, with $n \geqslant 4$."

$$\log_2 n! \geqslant \frac{n}{2} \log_2 \frac{n}{2}$$

"So now with $n \geqslant 4$, since $\log_2 \frac{n}{2} \geqslant \frac{1}{2} \log_2 n$, we get this."

$\log_2 n! \geqslant \frac{n}{2} \log_2 \frac{n}{2}$

$\geqslant \frac{n}{2} \frac{1}{2} \log_2 n$ because $\log_2 \frac{n}{2} \geqslant \frac{1}{2} \log_2 n$

$= \frac{1}{4} n \log_2 n$

"And from that, we get this inequality..."

$$\log_2 n! \geqslant \frac{1}{4} n \log_2 n \quad (n \geqslant 4)$$

"...which we can rewrite using big-Omega notation like this."

$$\log_2 n! = \Omega(n \log_2 n)$$

"In other words..."

$$\log n! = \Omega(n \log n)$$

"And that is what we wanted to prove. Done and done."

> **Answer 6-3 (Maximum number of comparisons in a comparative sort)**
>
> By evaluating the height of a comparative tree, we can say that the maximum number of comparisons in a comparative sort of an array with n elements is
>
> $$T_{\max}(n) = \Omega(n \log n).$$

Tetra groaned as she tried to keep up in her note-taking.

"Hold up, I'm stuck on $\log_2 \frac{n}{2} \geqslant \frac{1}{2} \log_2 n$ for $n \geqslant 4$."

"Like we did before, just look at the sign of their difference," I said.

$$\begin{aligned}
\log_2 \frac{n}{2} - \frac{1}{2} \log_2 n &= (\log_2 n - \log_2 2) - \frac{1}{2} \log_2 n &&\text{def. of logarithms} \\
&= \frac{1}{2} \log_2 n - \log_2 2 &&\text{combine } \log_2 n \text{ terms} \\
&= \frac{1}{2} \log_2 n - 1 &&\text{because } \log_2 2 = 1
\end{aligned}$$

"If $n \geqslant 4 = 2^2$, then $\log_2 n \geqslant 2$, so we get this."

$$\frac{1}{2} \log_2 n - 1 \geqslant \frac{1}{2} \cdot 2 - 1 = 0$$

"This means $\log_2 \frac{n}{2} - \frac{1}{2} \log_2 n \geqslant 0$, so we end up with this."

$$\log_2 \frac{n}{2} \geqslant \frac{1}{2} \log_2 n$$

"Got it, thanks! Looks like I need to bone up on my logarithms."

"So with all this we've shown that when implementing a comparative sort, we're going to need to perform at least an order of $n \log n$ comparisons," I said.

Miruka nodded. "What's really interesting, though, is that we were able to do the proof without explicitly saying anything about how we're going to implement the algorithm. We just focus on the 'compare' operation, and evaluate the static structure of a comparison tree." She paused to reflect on that before continuing. "So we've used a comparison tree to show that the maximum number of comparisons in a comparative *sort* will be of order at least $n \log n$. But we can do something similar to show that the maximum number of comparisons in a comparative *search*, in other words an algorithm in which we search for an element in an array using only comparisons, is $\Omega(\log n)$. The maximum number of comparisons in a binary search was $O(\log n)$. In other words, a binary search is $\Theta(\log n)$, which makes it asymptotically the best algorithm. Meaning no matter how hard you try, you could never come up with a comparative search that is asymptotically faster than a binary search."

"Huh," I said.

"It's interesting that trees keep showing up," Tetra said, "like in the comparison tree you used just now, and in the tree diagrams that pop up in number-of-cases problems. I love it when similar concepts pop up in unexpected places."

"Good point," I said. "I guess trees just have a structure that's good for organizing information."

6.6 Passing On and Learning

6.6.1 Passing On

Tetra's expression turned serious.

"When we want a computer to actually do these things, we have to write a program to tell it what to do, right? So it's like we're passing something on to the computer, just like we can pass on math to other people."

"I get that," I said.

Tetra glanced at me and gave a quick smile, but her face soon became solemn again.

"You taught me that mathematics is a message, one we receive from the past through books and texts. But someday... someday I want to be the sender of those messages, not just a recipient. And I'm finally getting to the point where I think that day might actually come."

"Wow, I was thinking pretty much the exact same thing just recently. How I want to send those messages, not just solve problems that somebody else has already solved."

"You too?" she said. After a pause to think, she continued. "When I think about math, I imagine this far-off world."

"Far off?"

"In the past, yes. So what I want to pass on is something to some far off future world, one in which I won't be here any longer, not physically at least."

Speechless, I noticed Tetra was holding clenched fists in front of her chest.

Wow, big dreams in such a small package.

I sat silently for a moment, wondering just how I might pass messages to the future, a future far beyond my own short life.

"Papers," Miruka said.

"Papers?"

"Papers, as in academic papers?" Tetra said. "You're saying that's how we should pass on our messages? Sounds like a bit much for me."

"Papers aren't about writing something that's difficult. They're about writing something worthy of passing on in way that it will be correctly understood. Research is about adding your own discoveries to the other discoveries of humanity. Learning is about stacking the now on top of the past so that we can see the future. It's about standing on the shoulders of giants."

6.6.2 Learning

"Where did you learn everything you know?" Tetra asked.

"From books and papers, of course, not to mention my teachers," Miruka replied.

"Teachers? I don't think I've ever seen you studying the kind of math they teach us at school."

"Not just school teachers. Lisa's mother, Dr. Narabikura, taught me a lot too."

"Huh."

"I go to a lot of seminars at the Narabikura Library. Dr. Narabikura hosts one every time she's back visiting from the States. Lots of good stuff there. Interesting problems and how to solve them, the essence of learning, how to approach books and papers... She's taught me a lot."

"Not me," Lisa said, looking up from her computer and giving her red mane a shake. We all turned to look at her. "Never taught me anything. I didn't learn a single thing from her." She started fiercely coughing, and glared at Miruka.

"Well I sure did," Miruka deadpanned. "Never saw you at any of her study groups, though, despite you living right next to the library. Nobody's going to force you to participate in events taking place practically in your own back yard, but since you chose not to, I don't think you have any right to complain."

"I was there," Lisa said, then coughed again.

"At first, maybe, but not for long. I wonder why?"

"Because—"

"Enough. Doesn't matter," Miruka spat. "You can talk with Dr. Narabikura any time you want. She's your own mother, after all. Are you trying to say you asked her questions she refused to answer?"

Lisa sat in silence.

"So how dare you complain that your mother taught you nothing, when you did nothing to take advantage of the opportunities right in front of you?"

"Because—" *cough* "Because I wasn't honest with myself." *cough* "Can't talk well..." *cough* "I missed my chance, so it's too late. Now, I'm on my own."

"Hmph. Decide you've missed your chance if you like. Decide it's too late if you like. Remain a cynic if you like. Do whatever you want. Show everyone exactly how you approach learning. You don't want to solve mysteries, you just want to protect your own feelings."

Despite the heat of Miruka's words, her voice was icy cold.

"C'mon guys, let's not fight!" Tetra said, eyes flicking back and forth between Miruka and Lisa. "We're all students of Euler here, right? Right?" She hesitantly flipped a Fibonacci sign at the two, but their eyes remained locked on each other.

"Nobody's fighting," Miruka said, but her stare never left Lisa.

Lisa finally turned away, put on her headphones, sank back behind her computer, and tuned us out.

Verbally, $O(f(n))$ can be read as "order at most $f(n)$"; $\Omega(f(n))$ as "order at least $f(n)$"; $\Theta(f(n))$ as "order exactly $f(n)$."

DONALD KNUTH
Selected Papers on Analysis of Algorithms [8]

CHAPTER 7

Matrices

> I had no compass on board, and should never have known how to have steered towards the island, if I had but once lost sight of it.
>
> DANIEL DEFOE
> *Robinson Crusoe*

7.1 IN THE LIBRARY

7.1.1 *Ms. Mizutani*

Ms. Mizutani surprised Tetra and me like she never had before: "I need some help," she said.

We were in the library after classes had ended. Off balance from her unexpected approach, I glanced at my watch. Normally she appeared in the middle of the library like clockwork, always in a tight skirt and darkly tinted eyeglasses, to loudly announce that the library was closing and we had to go home. It was still too early for that, though.

"I need some help," she repeated.

I looked around, but Tetra and I were the only ones in the library. Which meant that she could only be speaking to us.

"Yes, ma'am!" Tetra said, raising her hand.

Ms. Mizutani gestured for us to follow, and we tentatively allowed her to lead us into the library's storage room—a first experience for us both. As it turned out, she needed help shelving some books that were to be stored there.

The storage area was a maze of shelves that reached to the ceiling. It felt, and somewhat smelled, like we were lost in a forest. I was reminded of the Narabikura Library, which combined the scents of books and the ocean.

After giving her instructions, Ms. Mizutani slipped back into her office. Tetra and I shrugged, and began moving books from racks to shelves.

7.1.2 Organic Tetras

Tetra passed me a book titled *Organic Chemistry*.

"Maybe I should study this instead of computer science," she said.

"Why? You seem to be doing great with algorithms."

"Because I'd be in good company. There are Tetras everywhere!"

"Huh?"

"Like, if you have a string of forty carbon atoms holding as many hydrogen atoms as they can, that's called tetracontane!"

"Ah, saturated hydrocarbons, right?"

"Right! And if you have one that's four hundred carbon atoms long, that's tetractane."

"There's a name for that?"

"It doesn't stop there! Four thousand carbon atoms and you get tetraliane!"

"But hang on, if there's just four carbon atoms, isn't that butane?"

"Yeah, but if I can win a Nobel prize in chemistry, I bet they'd let me rename it to tetrane."

"Good luck with that," I said, laughing.

After a few more minutes, we were done shelving books.

"Here's the last," Tetra said, handing me a linear algebra textbook. It was heavier than it looked, and slipped out of my fingers as I took it from her.

"Oops!" we both said, and both bent over to pick it up—
bonk
"Ouch!" "Owww!"

"Sorry about that," Tetra said, rubbing her forehead as she picked up the book.

"Nah, my fault."

"Are you okay?" Still holding the book in her left hand, Tetra stood on her toes and reached out with her right as if to rub my forehead. Her citrus scent seemed stronger than usual. Her eyes locked on mine, and mine on hers.

The room became silent and still, and I felt my hands rising and landing on her shoulders, as if drawn there.

Tetra pressed the book close to her chest with both arms, and tilted her head. "Um, what—?"

I said nothing, instead pulling her toward me, and—

boof

Tetra slammed the book into my chest, knocking me back.

"This is wrong," she said. She ran out of the room, leaving me with not her but *Linear Algebra* in my arms.

7.2 YURI

7.2.1 *Inconsistent*

"This is wrong?!" Yuri shouted. She was comparing her homework answers with those in the back of her textbook. "Oh. Yeah. I guess it is. Okay, done!"

It was Saturday and Yuri was in my room, as usual. I had told her I needed to study that day, but I'm not sure why I'd bothered. Of course she had showed up with an armful of homework.

She folded up her plastic glasses and stuck them in her breast pocket.

"Now I'm bored. Entertain me," she said.

I looked up from my world history textbook and sighed.

"I told you I have to study," I said, but I already saw where this was going. "Here, play with some simultaneous equations."

I scribbled a couple of lines on a piece of paper and passed it to Yuri.

$$\begin{cases} 2x + 4y = 7 & \cdots \text{\textcircled{1}} \\ x + 2y = 4 & \cdots \text{\textcircled{2}} \end{cases}$$

"Aw, come on," she said. "Too easy. Just multiply both sides of ② by 2 to get rid of the x. That gives you a 2x to work with. Call it ③."

$$2x + 4y = 8 \quad \cdots \text{ ③ (doubled both sides of ②)}$$

"Work with 2x how?" I said.

"By subtracting ① from ③ to make x go poof."

$$\begin{array}{r} 2x + 4y = 8 \quad \cdots \text{ ③} \\ -\quad 2x + 4y = 7 \quad \cdots \text{ ①} \\ \hline 0 + 0 = 1 \; (?) \end{array}$$

"Oh yeah?" I said.

"What happened?" Yuri said. "Where'd my y's go? What's this $0 + 0 = 1$? Something's broke!"

"You broke math?"

"No, you gave me broken math! Check it out. ① says that $2x+4y$ is 7, but ③ says $2x+4y$ is 8. No way you can find an x and a y like that."

"Okay, you got me. There's no solution for this system of equations. It's inconsistent."

"Inconsistent?"

"Right. Like you said, there's no pair of x and y values that can make both equations true."

7.2.2 Underdetermined

I pulled out another piece of paper and wrote a new set of equations.

"Here," I said. "In this one I've changed the 7 in ① to 8."

$$\begin{cases} 2x + 4y = 8 & \cdots \text{ⓐ} \\ x + 2y = 4 & \cdots \text{ⓑ} \end{cases}$$

Yuri nodded. "Okay, sure."

"So now let's multiply both sides of ⓑ by 2, like you did before."

$$2x + 4y = 8 \quad \cdots \text{ double both sides of ⓑ}$$

"Oops, now you have the same equation as ⓐ."

"That's right. So ⓐ and ⓑ are in essence one and the same. There's really only one equation in this system. Any x, y pair that satisfies ⓐ will also satisfy ⓑ, so (x, y) could be $(0, 2)$ or $(2, 1)$ or $(\frac{1}{2}, \frac{7}{4})$ or any number of other pairs."

"Right."

"A system of equations like this, one where there's infinitely many possible solutions, is called underdetermined."

"Inconsistent and underdetermined," Yuri said. "Huh."

Yuri sat quietly for a moment, slowly waving her chestnut ponytail back and forth, a sure sign she was in deep-think mode. I waited, watching the sun flash off her hair.

"Check it out," she finally said. "You made an inconsistent system of equations, and an underdetermined one, but if you have a normal system of equations there's just one solution, right? So what makes that happen?"

"An excellent question," I said.

"How so?" she asked. Yuri was a big fan of logic, the feeling of closing in on an answer.

"A system of equations with just one solution is called regular."

"Regular? I don't think I learned that in school."

"You probably wouldn't, unless you got deep into the conditions for finding a solution to a system of equations."

"Fair enough. Okay, so if you can multiply these numbers 2 and 4 here—"

"Coefficients," I said. "Call them coefficients."

$$\underline{2}x + \underline{4}y = 7 \qquad \text{(coefficients)}$$

"Fine. So if you can multiply these *coefficients* by some number and you end up with the same as the other ones—"

"Hang on, Yuri. I know you understand this, but with an explanation like that you'll never get someone else to understand it. Actually, if you can't explain something in a way that someone new to it can understand, you can't really say you understand it yourself."

"Yeah, but still..." Yuri tapped a tooth while she thought.

After a time, I said, "Okay, so let's talk about the conditions for a system of equations to be regular."

7.2.3 Regular

> **Problem 7-1 (conditions for regularity in a system of equations)**
>
> Find the conditions under which the system of equations
>
> $$\begin{cases} ax + by = s \\ cx + dy = t \end{cases}$$
>
> will have a unique solution.

"This is a generalized system of two equations with coefficients a, b, c, d," I said. "See if you can use this to give a proper explanation of the conditions under which

$$\begin{cases} ax + by = s \\ cx + dy = t \end{cases}$$

will have a unique solution."

"Mmm, let me do it using ①② and ⓐⓑ instead."

"Sure, go ahead."

$$\text{Inconsistent—} \begin{cases} 2x + 4y = 7 & \cdots \text{①} \\ x + 2y = 4 & \cdots \text{②} \end{cases}$$

$$\text{Underdetermined—} \begin{cases} 2x + 4y = 8 & \cdots \text{ⓐ} \\ x + 2y = 4 & \cdots \text{ⓑ} \end{cases}$$

"Since $x + 2y$ is the same as $1x + 2y$, I guess the coefficient for x is 1? So if we halve the coefficients for ① and ⓐ, which are 2 and 4, we get the coefficients for ② and ⓑ, which are 1 and 2. That's when we get an inconsistent or underdetermined system of equations."

"Okay," I said. "You're getting there. But since the coefficients in the problem are written as a, b, c, d, why not just leave them that way?"

"Leave them what way how?"

"Like this. Here's where we start."

$$\begin{cases} ax + by = s & \cdots \text{\textcircled{A}} \\ cx + dy = t & \cdots \text{\textcircled{B}} \end{cases}$$

"Let's calculate Ⓐ × d − b × Ⓑ to get rid of the y. First, we let Ⓒ be Ⓐ × d ..."

$$adx + bdy = sd \qquad \cdots \text{\textcircled{C}}$$

"... then let Ⓓ be b × Ⓑ."

$$bcx + bdy = bt \qquad \cdots \text{\textcircled{D}}$$

"Now when we calculate Ⓒ − Ⓓ, the y's go away."

$$\begin{array}{r} adx + bdy = sd \quad \cdots \text{\textcircled{A}} \times d \\ -\quad bcx + bdy = \quad bt \cdots b \times \text{\textcircled{B}} \\ \hline (ad - bc)x \quad = sd - bt \end{array}$$

"So if $ad - bc \neq 0$, we can find x like this."

$(ad - bc)x = sd - bt$ from the above calculation

$x = \dfrac{sd - bt}{ad - bc}$ divide both sides by $ad - bc$

"Okay, so next—"

"I know!" Yuri said. "We've found x, so now we plug that in to Ⓐ and solve for y, right?"

"You can do that, sure. Or we can look carefully at the equation and do the same thing to x to get rid of the y."

"Huh? How's that?"

"Like this."

$$\begin{cases} ax + by = s & \cdots \text{\textcircled{A}} \\ cx + dy = t & \cdots \text{\textcircled{B}} \end{cases}$$

"This time we want to get rid of x," I said, "so we calculate $a \times \text{\textcircled{B}} - \text{\textcircled{A}} \times c$."

$$\begin{array}{r} acx + ady = at \quad \cdots a \times \text{\textcircled{B}} \\ -\quad acx + bcy = \quad sc \cdots \text{\textcircled{A}} \times c \\ \hline (ad - bc)y = at - sc \end{array}$$

"Once again, if $ad - bc \neq 0$, we can find y like this."

$$(ad - bc)y = at - sc \quad \text{from the above calculation}$$

$$y = \frac{at - sc}{ad - bc} \quad \text{divide both sides by } ad - bc$$

"From this, we can say that if $ad - bc \neq 0$, then the system of equations has only this unique solution."

$$x = \frac{sd - bt}{ad - bc}, \ y = \frac{at - sc}{ad - bc}$$

"So the only condition we have is this."

$$ad - bc \neq 0$$

"Got it?" I asked, looking up at Yuri.

"I dunno," she said. "It feels like you skipped something somewhere."

Yuri started playing with her hair as she thought. I waited silently, thinking about how much her attitude toward math had changed. It wasn't long ago that she would have just given up, saying that she didn't understand, but recently she was showing real determination. I wondered if something of Tetra was rubbing off on her.

"Okay," Yuri said, "I get that there's a unique solution if $ad-bc \neq 0$, but it feels like we still haven't proved there's *no* unique solution if $ad - bc = 0$. Seems like there's something missing without doing that."

"Nicely spotted, Yuri. Okay, so let's see if we can say that there is no unique solution when $ad - bc = 0$, going back to the first system of equations."

I stopped there, however, realizing that the manipulations of the equations I would have to do might be a bit much for Yuri. Then I remembered something that Miruka had told me many times: "Your problem," she often said, "is that you never draw graphs."

"Actually," I said, "you know that if we plot on a graph every (x, y) point that satisfies the equation $ax + by = s$, we get a line, right?"

"Now I do."

"In fact, $ax + by = s$ and $cx + dy = t$ will both be straight lines. We can use graphs of those lines to think about a system of equations, which I think will give you a better idea of what we're doing. First, we need to graph $ax + by = s$."

"Sounds good."

"First, we can move things around a bit to make it easier to see what the line looks like."

$$ax + by = s \qquad \text{one of the equations}$$
$$by = -ax + s \qquad \text{move } ax \text{ to the right}$$
$$y = \underbrace{-\frac{a}{b}}_{\text{slope}} x + \underbrace{\frac{s}{b}}_{\text{y-intercept}} \qquad \text{assumes } b \neq 0$$

"So the slope of $ax + by = s$ is $-\frac{a}{b}$, and the y-intercept, the value where the line crosses the y-axis, is $\frac{s}{b}$. You see that from the graph, right?"

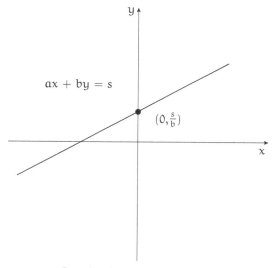

Graph of line $ax + by = s$.

"Mmm..."

"Do you see how a set of points (x, y) on a plane can form a geometric figure? How here all points that make the equation $ax + by = s$ true form a straight line?"

"Yeah, yeah, I know all that. What I want to know now is how you can just assume that $b \neq 0$ here. Things like that make me... itchy."

"Well let me see if I can scratch that itch, by taking a look at what kind of shape $ax + by = s$ can describe."

"Scratch away!"

▶ *When* $a = 0 \wedge b = 0 \wedge s = 0$

$$0x + 0y = 0$$

"This will be true for any (x, y) you can think of, right? So in this case we're describing *the entire plane*."

▶ *When* $a = 0 \wedge b = 0 \wedge s \neq 0$

$$0x + 0y = s \qquad (s \neq 0)$$

"This can't be true for any (x, y), since there would be a zero on the left side, and some nonzero value on the right. So in this case, there is no shape at all that describes the $ax + by = s$ equation. We can also say that the set of points satisfying the equation is the null set."

▶ *When* $a = 0 \wedge b \neq 0$

$$0x + by = s$$

"In this case $b \neq 0$, so we can write this."

$$y = \frac{s}{b}$$

"Here, it doesn't matter what the value of x is, we just need $y = \frac{s}{b}$. In other words, the shape that satisfies $ax + by = s$ will be a line parallel to the x-axis, a horizontal line that intersects the y-axis at $(0, \frac{s}{b})$."

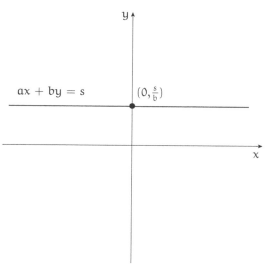

When $a = 0 \wedge b \neq 0$, $ax + by = s$ is a horizontal line.

▶ *When $a \neq 0 \wedge b = 0$*

$$ax + 0y = s$$

"Since $a \neq 0$, we get this."

$$x = \frac{s}{a}$$

"This case is a lot like the last one. We know that $x = \frac{s}{a}$, no matter what value y takes, which means the shape describing $ax + by = s$ is a line parallel to the y-axis, a vertical line intersecting the x-axis at $(\frac{s}{a}, 0)$."

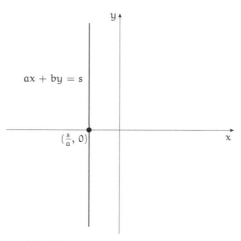

When $a \neq 0 \wedge b = 0$, $ax + by = s$ is a vertical line.

▶ *When $a \neq 0 \wedge b \neq 0$*
$$y = -\frac{a}{b}x + \frac{s}{b}$$

"This will be a straight line with slope $-\frac{a}{b}$ and y-intercept $(0, \frac{s}{b})$, specifically one that is slanted, not horizontal or vertical."

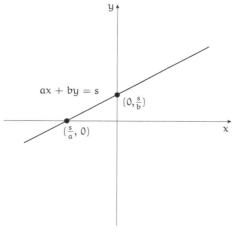

When $a \neq 0 \wedge b \neq 0$, $ax + by = s$ is a slanted line.

"Way too much work, dude," Yuri said. "Not that it's not interesting, mind you. It's kinda cool to see how the equations and graphs are related."

"We could also do a case-by-case for $cx + dy = t$, but we'd just be repeating the same thing with different letters, so let's just summarize everything in a table. I don't know a good word to describe a nonexistent shape, so I'll just use 'null set' for that."

	$ax + by = s$	$cx + dy = t$
Full plane	$a = 0 \wedge b = 0 \wedge s = 0$	$c = 0 \wedge d = 0 \wedge t = 0$
Null set	$a = 0 \wedge b = 0 \wedge s \neq 0$	$c = 0 \wedge d = 0 \wedge t \neq 0$
Horiz. line	$a = 0 \wedge b \neq 0$	$c = 0 \wedge d \neq 0$
Vert. line	$a \neq 0 \wedge b = 0$	$c \neq 0 \wedge d = 0$
Slanted line	$a \neq 0 \wedge b \neq 0$	$c \neq 0 \wedge d \neq 0$

"Now this I like," Yuri said.

"Any solution to the system of equations that we're looking at will be an (x, y) pair that makes *both* of the equations true. Remembering that we're thinking of the forms of these equations as sets of points, this means the point has to be on *both* shapes that the equations describe."

"Oh ho! Interesting!"

"Okay, now look at this table, and think about all the pairs you could possibly have, like 'the entire plane and a horizontal line,' or 'a slanted line and a horizontal line.' But also, think about which pairs will share a single point in all cases. That will tell you what pairs of equations will give you a unique answer."

"Uh-oh, we're getting into too-much-work territory again."

"Aw c'mon, it's not all that bad. We can lump lots of cases together. Like, for the 'entire plane and horizontal line' case, the shared area will be the horizontal line itself, right? And you can do the same thing for the entire plane plus whatever—the common form will be the whatever."

"Okay, so the entire plane ones are easy."

"There's also some where we can't be sure, like 'a horizontal line and a horizontal line.' The set of shared points might be the null set, if the two are different lines, or it may be the line itself, if both are the same line."

"Aha! But a vertical line and a horizontal line will definitely have a single shared point, right?"

"You bet. See? You get this."

Yuri and I joined forces to create a table for each pair of forms.

	Full plane	Null set	Horiz. line	Vert. line	Slanted line
Full plane	Full plane	Null set	Horiz. line	Vert. line	Slanted line
Null set	Null set	Null set	Null set	Null set	Null set
Horiz. line	Horiz. line	Null set	Null/Horiz.	Point	Point
Vert. line	Vert. line	Null set	Point	Null/Vert.	Point
Slanted line	Slanted line	Null set	Point	Point	?

Shared areas between planar forms.

"Great! What's next?" Yuri said.

"Next is making a table of $ad - bc$ values. For example, say that $ax + by = s$ is a vertical line, so $a \neq 0 \wedge b = 0$, and $cx + dy = t$ is a horizontal line, so $c = 0 \wedge d \neq 0$. Then what can you say about the value of $ad - bc$?"

"Hmm. I know that $a \neq 0$ and $d \neq 0$, so ad can't be zero. And since $b = 0$ and $c = 0$, I know bc is definitely zero. So that means $ad - bc$ can't be zero!"

"Exactly right. So now we want a table that shows all of these zero–nonzero results."

"Gotcha!"

	Full plane	Null set	Horiz. line	Vert. line	Slanted line
Full plane	Zero	Zero	Zero	Zero	Zero
Null set	Zero	Zero	Zero	Zero	Zero
Horiz. line	Zero	Zero	Zero	Non-zero	Non-zero
Vert. line	Zero	Zero	Non-zero	Zero	Non-zero
Slanted line	Zero	Zero	Non-zero	Non-zero	?

Zero–nonzero values for $ad - bc$.

"Okay, compare these two tables," I said. "See how the places where the common part between forms is 'single point' and the places where the value of $ad - bc$ is 'nonzero' match up perfectly?"

"Oh, yeah! Look at that! Except for this 'slanted line and slanted line' place, though. What's with the question mark?"

"Well, think about what happens when both shapes are slanted lines, with $ax + by = s$ having a slope of $-\frac{a}{b}$ and $cx + dy = t$ having a slope of $-\frac{c}{d}$. What would the difference in those slopes be?"

" 'Difference' means I should be subtracting, right?"

$$\{\text{Slope of } ax + by = s\} - \{\text{slope of } cx + dy = t\}$$
$$= \left(-\frac{a}{b}\right) - \left(-\frac{c}{d}\right)$$
$$= -\frac{a}{b} + \frac{c}{d}$$
$$= -\frac{ad}{bd} + \frac{bc}{bd}$$
$$= -\frac{ad - bc}{bd}$$

"Great," I said. "See the $ad - bc$ there? What would it mean for that part to be zero?"

"Something like this, right?"

$$\{\text{Slope of } ax + by = s\} - \{\text{slope of } cx + dy = t\} = 0$$

"Sure. Which means we could also say this."

$$\{\text{Slope of } ax + by = s\} = \{\text{slope of } cx + dy = t\}$$

"In other words, if $ad - bc = 0$ then the two lines have the same slope, and $ad - bc \neq 0$ would mean they have different slopes. But if two slanted lines have only one point in common, they can't have the same slope, right?"

"Sure, I see that! If they're parallel they won't have any points in common, and if they're on top of each other they share all their points."

"Exactly. And that's everything we need to answer your question. If $ad - bc = 0$ there is no unique solution, and if $ad - bc \neq 0$ there is one. We know that has to be the case, because we looked at every possibility."

"Indeed we did."

"Summing up..."

$ad - bc \neq 0 \iff$ The system of equations has a unique solution

---

Answer 7-1 (conditions for regularity in a system of equations)

The system of equations

$$\begin{cases} ax + by = s \\ cx + dy = t \end{cases}$$

will have a unique solution if and only if

$$ad - bc \neq 0.$$

7.2.4 A Letter

Yuri grabbed a lemon candy off of my desk and popped it into her mouth.

"By the way, Tetra told me something *very* interesting," she said, causing my heart to skip a beat.

"Uh...yeah? What's that?" I managed.

"She told me I should always write using a pen, thinking hard while I write as neatly as possible, and to always express my true feelings. And that I should wait until after I'd finished writing to decide whether to actually send it. The most important thing is to use words to work out how you feel."

"What on earth are you going on about?"

"About writing letters, of course! I found out that guy is going to move, you see..."

That guy. Apparently Yuri had gone to Tetra for advice, and she had suggested writing a letter. Very Tetra-like advice indeed.

"Ah. Well."

"To tell you the truth I used to think Tetra was a bit ditzy, but I was wrong. She's a good friend to have."

"Hey, Yuri."

"Yeah?"

"How do you deal with it when things get, like, *weird* between you and a friend?"

"Well that's what I was just talking about, isn't it? You have to talk with them. Not using big words or cool words, just using honest words. Words that they will understand. Otherwise, what's the point?"

7.3 Tetra

7.3.1 In the Library

I went back to the library after classes on Monday. I walked toward Tetra, who was focused on the notebook she was writing in. She looked up, and I passed her a note.

Sorry about the other day.

She glanced at me, then added a line and handed it back.

No worries!

She looked up, and gave me a smile that did a lot to untangle the knot of awkwardness that was inside me. Her smile was another of her superpowers.

7.3.2 Rows and Columns

I glanced at Tetra's notes.

"Rotations using matrices?" I said.

"Nothing that advanced. I'm just not doing super well with matrices in general, so I was reviewing my notes."

Tetra had such a good sense about these things, a kind of unease she couldn't shake until she was sure she had truly understood something. Only when something had truly clicked would she move on.

"Honestly, I'm not even really sure what matrices are in the first place. I can do calculations on them like I learned in class, but I make so many mistakes..."

"Want to talk about them some?"

"Yeah, sure! Of course! But starting at the very beginning, if you don't mind."

"Well, I guess the very beginning would be the difference between rows and columns."

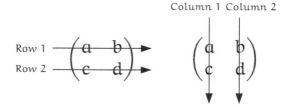

"Which I never fail to get confused," Tetra said.

"You wouldn't be the first. Try remembering it this way: in a movie theater, the seats are arranged in left-to-right *rows*, while *columns* go up–down, like the columns in ancient Greek architecture."

"Perfect! Now I'll never get them confused again!"

7.3.3 Matrix–Vector Products

"You've learned how to multiply a matrix by a vector, right?" I asked.

$$\begin{pmatrix} a & b \\ c & d \end{pmatrix} \begin{pmatrix} x \\ y \end{pmatrix} = \begin{pmatrix} ax + by \\ cx + dy \end{pmatrix}$$

"I have, more or less. But I still make mistakes all the time."

"Here's how to remember. You want $ax + by$ and $cx + dy$ on the right, right?"

"Sure."

"You get that by multiplying a and x, then multiplying b and y, then adding ax and by together. You just need to remember that sum of products: multiply-multiply-add, multiply-multiply-add."

$$\underbrace{\overbrace{ax}^{\text{multiply}} + \overbrace{by}^{\text{multiply}}}_{\text{add}}$$

"Multiply-multiply-add. Got it!" Tetra said.
"Twice, though. Here's the first one."

$$\begin{pmatrix} a & \cdot \\ \cdot & \cdot \end{pmatrix} \begin{pmatrix} x \\ \cdot \end{pmatrix} = \begin{pmatrix} ax \cdots \cdots \\ \cdot \end{pmatrix} \qquad \text{multiply}\ldots$$

$$\begin{pmatrix} \cdot & b \\ \cdot & \cdot \end{pmatrix} \begin{pmatrix} \cdot \\ y \end{pmatrix} = \begin{pmatrix} \cdots \cdots by \\ \cdot \end{pmatrix} \qquad \text{multiply}\ldots$$

$$\begin{pmatrix} a & b \\ \cdot & \cdot \end{pmatrix} \begin{pmatrix} x \\ y \end{pmatrix} = \begin{pmatrix} ax + by \\ \cdot \end{pmatrix} \qquad \text{add.}$$

"Oh, neat!" Tetra said. "Let me try the second one!"

$$\begin{pmatrix} \cdot & \cdot \\ c & \cdot \end{pmatrix} \begin{pmatrix} x \\ \cdot \end{pmatrix} = \begin{pmatrix} \cdot \\ cx \cdots \cdots \end{pmatrix} \qquad \text{multiply}\ldots$$

$$\begin{pmatrix} \cdot & \cdot \\ \cdot & d \end{pmatrix} \begin{pmatrix} \cdot \\ y \end{pmatrix} = \begin{pmatrix} \cdot \\ \cdots \cdots dy \end{pmatrix} \qquad \text{multiply}\ldots$$

$$\begin{pmatrix} \cdot & \cdot \\ c & d \end{pmatrix} \begin{pmatrix} x \\ y \end{pmatrix} = \begin{pmatrix} \cdot \\ cx + dy \end{pmatrix} \qquad \text{add.}$$

"There ya go. Pull them together, and we're done."

$$\begin{pmatrix} a & b \\ c & d \end{pmatrix} \begin{pmatrix} x \\ y \end{pmatrix} = \begin{pmatrix} ax + by \\ cx + dy \end{pmatrix}$$

"I think I've got it, but let me practice a little."

Tetra went back to her notebook, and worked through several problems.

"Okay, much better," she said. "This is the pattern, right?"

$$\begin{pmatrix} \Rightarrow & \Rightarrow \\ \cdot & \cdot \end{pmatrix} \begin{pmatrix} \Downarrow \\ \Downarrow \end{pmatrix} = \begin{pmatrix} ax + by \\ \cdot \end{pmatrix}$$

$$\begin{pmatrix} \cdot & \cdot \\ \Rightarrow & \Rightarrow \end{pmatrix} \begin{pmatrix} \Downarrow \\ \Downarrow \end{pmatrix} = \begin{pmatrix} \cdot \\ cx + dy \end{pmatrix}$$

"I don't think I ever would have gotten this from just staring at my textbook," she said. "Actually writing it out, multiplying this and this, adding that and that makes all the difference."

"I agree completely," I said, nodding. "Writing things out by hand can show you patterns you wouldn't have noticed otherwise."

"Especially when you're learning something new like this."

"Except that this isn't really new, you know. You saw it in junior high school."

"Matrices? In junior high? I don't think so..."

"Sure, though they probably weren't called matrices."

"What were they called, then?"

"Systems of equations."

7.3.4 Systems of Equations and Matrices

"A system of equations like this, for example," I said.

$$\begin{cases} 3x + y = 7 \\ x + 2y = 4 \end{cases}$$

"Now *this* I definitely saw in junior high," Tetra said. "Are we going to solve it?"

"Let's take a look at these equations first."

$$3x + y = 7$$

"Do you see the pattern hidden in there?"

"Um..."

"Maybe it's easier to see like this."

$$\underbrace{\overbrace{3 \cdot x}^{\text{multiply}} + \overbrace{1 \cdot y}^{\text{multiply}}}_{\text{add}} = 7$$

"Check it out! It's our friend, multiply-multiply-add! A sum of products!"

"That's right. So we can use a matrix to represent this system of equations."

$$\begin{cases} 3x + y = 7 \\ x + 2y = 4 \end{cases} \quad \longleftrightarrow \quad \begin{pmatrix} 3 & 1 \\ 1 & 2 \end{pmatrix} \begin{pmatrix} x \\ y \end{pmatrix} = \begin{pmatrix} 7 \\ 4 \end{pmatrix}$$

"Oh cool! You've rewritten the system of equations as a matrix!"

"Just one specific system of equations here, but we can generalize this to better show the relation between simultaneous equations and matrices."

$$\begin{cases} ax + by = s \\ cx + dy = t \end{cases} \quad \longleftrightarrow \quad \begin{pmatrix} a & b \\ c & d \end{pmatrix} \begin{pmatrix} x \\ y \end{pmatrix} = \begin{pmatrix} s \\ t \end{pmatrix}$$

"Right! So a system of equations is the product of a matrix and a vector!"

7.3.5 Matrix Products

I continued talking.

"A matrix like $\begin{pmatrix} a & b \\ c & d \end{pmatrix}$ with two rows and two columns is called a 2×2 matrix. You can also call a matrix with two rows and one column like $\begin{pmatrix} x \\ y \end{pmatrix}$ a 2×1 matrix, or you can call it a column vector. You write it in the same way as $\binom{n}{k}$ when talking about combinations, but that's a completely different thing. The contexts are completely different too, though, so you usually won't get confused between the two."

$\begin{pmatrix} a & b \\ c & d \end{pmatrix}$ a 2×2 matrix (second-order square matrix)

$\begin{pmatrix} a \\ b \end{pmatrix}$ 2×1 matrix (column vector)

$\begin{pmatrix} a & b \end{pmatrix}$ 1×2 matrix (row vector)

"Got it," Tetra said.

"Once you get used to that multiply-multiply-add pattern, you shouldn't have any problem moving from multiplying matrices and vectors to multiplying two matrices."

$$\begin{pmatrix} a & b \\ c & d \end{pmatrix} \begin{pmatrix} x & s \\ y & t \end{pmatrix} = \begin{pmatrix} ax+by & as+bt \\ cx+dy & cs+dt \end{pmatrix}$$

"Hmm... Something like this?"

$$\begin{cases} \begin{pmatrix} a & b \\ c & d \end{pmatrix} \begin{pmatrix} x \\ y \end{pmatrix} = \begin{pmatrix} ax+by \\ cx+dy \end{pmatrix} \\ \begin{pmatrix} a & b \\ c & d \end{pmatrix} \begin{pmatrix} s \\ t \end{pmatrix} = \begin{pmatrix} as+bt \\ cs+dt \end{pmatrix} \end{cases} \longleftrightarrow \begin{pmatrix} a & b \\ c & d \end{pmatrix} \begin{pmatrix} x & s \\ y & t \end{pmatrix} = \begin{pmatrix} ax+by & as+bt \\ cx+dy & cs+dt \end{pmatrix}$$

"Nice! You're definitely getting a feel for this."

"I guess. All these letters make me a bit dizzy, but I guess it's just a matter of carefully plugging away, one step at a time."

"It is. Keeping the pattern in mind, of course."

"Right. Multiply-multiply-add, multiply-multiply-add..."

Tetra looked back at her notes, her head slightly bobbing as she traced through a matrix.

"By the way," I said, "that multiply-multiply-add has a name: the inner product."

"Inner product..." Tetra said, writing in her notebook.

"For example, the inner product of two vectors $(a_1 \; a_2)$ and $\binom{b_1}{b_2}$ would be this."

$$a_1 b_1 + a_2 b_2$$

"See the multiply-multiply-add pattern? Multiplying matrices involves lots of these inner products."

"And it's quicker to say than multiply-multiply-add!"

7.3.6 Invertible matrices

"Okay, next let's think about how we can use matrices to solve systems of equations," I said.

"Okay, sure," Tetra said, nodding hesitantly.

"A minute ago, we represented a system of equations as a matrix and a vector, right?"

$$\begin{pmatrix} a & b \\ c & d \end{pmatrix} \begin{pmatrix} x \\ y \end{pmatrix} = \begin{pmatrix} s \\ t \end{pmatrix}$$

"Sure."

"So if we can get a system of equations into a form like this, then we've solved it."

$$\begin{pmatrix} x \\ y \end{pmatrix} = \begin{pmatrix} \cdots \\ \cdots \end{pmatrix}$$

"Oh, right, because then we'll know what the x and y values are!"

"Then let's give it a try." Having a flashback of Saturday with Yuri, I wrote out a generalized system of equations.

$$\begin{cases} ax + by = s & \cdots \text{\textcircled{A}} \\ cx + dy = t & \cdots \text{\textcircled{B}} \end{cases}$$

"Here we want to calculate $\text{\textcircled{A}} \times d - b \times \text{\textcircled{B}}$ so we can find x. In that calculation, we want to pay attention to what happens on the left side of $\text{\textcircled{A}}$ and $\text{\textcircled{B}}$."

$$\begin{aligned}
(\text{\textcircled{A}} \text{ left}) \times d - b \times (\text{\textcircled{B}} \text{ left}) &= d \times (\text{\textcircled{A}} \text{ left}) + (-b) \times (\text{\textcircled{B}} \text{ left}) \\
&= d(ax + by) + (-b)(cx + dy) \\
&= dax + dby + (-b)cx + (-b)dy \\
&= \bigl(da + (-b)c\bigr)x + \bigl(db + (-b)d\bigr)y \\
&= \bigl(da + (-b)c\bigr)x + \underbrace{\bigl(db + (-b)d\bigr)}_{\text{becomes 0}}y \\
&= \bigl(da + (-b)c\bigr)x
\end{aligned}$$

"Look! There it is!" Tetra said. "Multiply-multiply-add!"

$$\underbrace{\overbrace{da}^{\text{multiply}} + \overbrace{(-b)c}^{\text{multiply}}}_{\text{add}}$$

"That's the inner product of $(d \ -b)$ and $\binom{a}{c}$. Now let's watch the left side of $a \times \text{\textcircled{B}} - \text{\textcircled{A}} \times c$ to find y."

$$\begin{aligned}
a \times (\text{\textcircled{B}} \text{ left}) - (\text{\textcircled{A}} \text{ left}) \times c &= (-c) \times (\text{left side of } (\text{\textcircled{A}})) + a \times (\text{\textcircled{B}} \text{ left}) \\
&= (-c)(ax + by) + a(cx + dy) \\
&= (-c)ax + (-c)by + acx + ady \\
&= \bigl(\underbrace{(-c)a + ac}_{\text{becomes 0}}\bigr)x + \bigl((-c)b + ad\bigr)y \\
&= \bigl(\underbrace{(-c)b + ad}\bigr)y
\end{aligned}$$

"Another multiply-multiply-add here, the inner product of $(-c \ a)$ and $\binom{b}{d}$."

$$\underbrace{\overbrace{(-c)b}^{\text{multiply}} + \overbrace{ad}^{\text{multiply}}}_{\text{add}}$$

"Do these two inner products make you think of the matrix $\begin{pmatrix} d & -b \\ -c & a \end{pmatrix}$?"

$$\begin{cases} \text{inner product of } (d \ -b) \text{ and } \binom{a}{c} \\ \text{inner product of } (-c \ a) \text{ and } \binom{b}{d} \end{cases} \longleftrightarrow \begin{pmatrix} d & -b \\ -c & a \end{pmatrix} \begin{pmatrix} a & b \\ c & d \end{pmatrix}$$

"Now let's calculate the matrix product $\begin{pmatrix} d & -b \\ -c & a \end{pmatrix}\begin{pmatrix} a & b \\ c & d \end{pmatrix}$."

$$\begin{pmatrix} d & -b \\ -c & a \end{pmatrix}\begin{pmatrix} a & b \\ c & d \end{pmatrix} = \begin{pmatrix} da - bc & db - bd \\ -ca + ac & -cb + ad \end{pmatrix}$$
$$= \begin{pmatrix} ad - bc & 0 \\ 0 & ad - bc \end{pmatrix}$$
$$= (ad - bc)\begin{pmatrix} 1 & 0 \\ 0 & 1 \end{pmatrix}$$

"So assuming that $ad - bc \neq 0$, this should hold."

$$\frac{1}{ad - bc}\begin{pmatrix} d & -b \\ -c & a \end{pmatrix}\begin{pmatrix} a & b \\ c & d \end{pmatrix} = \begin{pmatrix} 1 & 0 \\ 0 & 1 \end{pmatrix}$$

"This matrix $\frac{1}{ad-bc}\begin{pmatrix} d & -b \\ -c & a \end{pmatrix}$ that we created is called the inverse of matrix $\begin{pmatrix} a & b \\ c & d \end{pmatrix}$."

"Hang on, hang on," Tetra said, holding up a hand. "You're losing me here. Weren't we trying to solve a system of equations? It seems like we've gotten away from that, sidetracked by all these strange calculations."

"Not at all. In fact, the inverse of the matrix is key to solving its system of equations. We represented the system of equations like this, right?"

$$\begin{pmatrix} a & b \\ c & d \end{pmatrix}\begin{pmatrix} x \\ y \end{pmatrix} = \begin{pmatrix} s \\ t \end{pmatrix}$$

"So let's multiply both sides by the inverse of the matrix, from the left."

$$\begin{pmatrix} a & b \\ c & d \end{pmatrix} \begin{pmatrix} x \\ y \end{pmatrix} = \begin{pmatrix} s \\ t \end{pmatrix}$$

$$\underbrace{\frac{1}{ad-bc} \begin{pmatrix} d & -b \\ -c & a \end{pmatrix} \begin{pmatrix} a & b \\ c & d \end{pmatrix}}_{} \begin{pmatrix} x \\ y \end{pmatrix} = \underbrace{\frac{1}{ad-bc} \begin{pmatrix} d & -b \\ -c & a \end{pmatrix}}_{} \begin{pmatrix} s \\ t \end{pmatrix}$$

$$\frac{1}{ad-bc} \begin{pmatrix} ad-bc & 0 \\ 0 & ad-bc \end{pmatrix} \begin{pmatrix} x \\ y \end{pmatrix} = \frac{1}{ad-bc} \begin{pmatrix} d & -b \\ -c & a \end{pmatrix} \begin{pmatrix} s \\ t \end{pmatrix}$$

$$\begin{pmatrix} 1 & 0 \\ 0 & 1 \end{pmatrix} \begin{pmatrix} x \\ y \end{pmatrix} = \frac{1}{ad-bc} \begin{pmatrix} d & -b \\ -c & a \end{pmatrix} \begin{pmatrix} s \\ t \end{pmatrix}$$

$$\begin{pmatrix} 1 & 0 \\ 0 & 1 \end{pmatrix} \begin{pmatrix} x \\ y \end{pmatrix} = \frac{1}{ad-bc} \begin{pmatrix} sd-bt \\ at-sc \end{pmatrix}$$

$$\begin{pmatrix} x \\ y \end{pmatrix} = \frac{1}{ad-bc} \begin{pmatrix} sd-bt \\ at-sc \end{pmatrix}$$

"Check it out!" Tetra said. "We got $\begin{pmatrix} x \\ y \end{pmatrix} = \cdots$, which means we've solved the system of equations!"

"Yep. And we did it by multiplying the system of equations by the inverse of its matrix."

"Sure is a complicated answer, though..."

"Don't let that hold you back. It's not so bad if you see some common patterns."

"Like what?"

"Like $ad - bc$ and $sd - bt$ and $at - sc$. Here, $ad - bc$ is called the determinant of the matrix $\begin{pmatrix} a & b \\ c & d \end{pmatrix}$. We write it as $\begin{vmatrix} a & b \\ c & d \end{vmatrix}$."

$$\begin{vmatrix} a & b \\ c & d \end{vmatrix} = ad - bc$$

$$\begin{vmatrix} s & b \\ t & d \end{vmatrix} = sd - bt$$

$$\begin{vmatrix} a & s \\ c & t \end{vmatrix} = at - sc$$

I paused to let Tetra write this in her notes, then continued.

"We can use the determinant to easily find the solution to a system of equations."

$$\begin{pmatrix} x \\ y \end{pmatrix} = \frac{1}{ad-bc} \begin{pmatrix} sd-bt \\ at-sc \end{pmatrix} \Leftrightarrow \begin{pmatrix} x \\ y \end{pmatrix} = \frac{1}{\begin{vmatrix} a & b \\ c & d \end{vmatrix}} \begin{pmatrix} \begin{vmatrix} s & b \\ t & d \end{vmatrix} \\ \begin{vmatrix} a & s \\ c & t \end{vmatrix} \end{pmatrix}$$

"That's what you call simple? Sheesh..."

"I'm just replacing part of the determinant $\begin{vmatrix} a & b \\ c & d \end{vmatrix}$ with s and t."

$$\begin{pmatrix} x \\ y \end{pmatrix} = \frac{1}{\begin{vmatrix} a & b \\ c & d \end{vmatrix}} \begin{pmatrix} \begin{vmatrix} \boxed{s} & b \\ \boxed{t} & d \end{vmatrix} \\ \begin{vmatrix} a & \boxed{s} \\ c & \boxed{t} \end{vmatrix} \end{pmatrix}$$

"Wow... I'm still not totally caught up with matrices and determinants, but I can certainly see there's lots of interesting stuff packed in here. You're right, I can't let the number of variables get me down. I'll just have to do my best to find patterns in all these inner products and determinants and equations and everything!"

7.4 Miruka

7.4.1 Finding Hidden Mysteries

I went back to the library again the following day, and found Tetra already there, talking to Miruka. Lisa was there too, for the first time in a while. She was sitting alone, though, near a window far from the others and (of course) typing on her computer. I figured she and Miruka still hadn't worked out whatever was going on between them.

"New card from Mr. Muraki," Miruka said when I approached.

Problem 7-2 (matrix exponentials)

$$\begin{pmatrix} 1 & 1 \\ 1 & 0 \end{pmatrix}^{10}$$

"He wants us to find the tenth power of this matrix?" I said.

"It was fun," Miruka said.

"And of course you've already done it," I said, rolling my eyes. *Probably in your head on your way here...* I mentally added.

"Don't tell me the answer!" Tetra said. She was already scribbling away in her notebook, working on the problem. I opened my own notebook and joined her.

The tenth power of this matrix. In other words...

$$\underbrace{\begin{pmatrix} 1 & 1 \\ 1 & 0 \end{pmatrix} \begin{pmatrix} 1 & 1 \\ 1 & 0 \end{pmatrix} \begin{pmatrix} 1 & 1 \\ 1 & 0 \end{pmatrix} \cdots \begin{pmatrix} 1 & 1 \\ 1 & 0 \end{pmatrix}}_{\text{ten of these}}$$

Well, I guess the direct approach would be to calculate the second power, then the third and so on, and look for patterns.

$$\begin{pmatrix} 1 & 1 \\ 1 & 0 \end{pmatrix}^1 = \begin{pmatrix} 1 & 1 \\ 1 & 0 \end{pmatrix}$$

$$\begin{pmatrix} 1 & 1 \\ 1 & 0 \end{pmatrix}^2 = \begin{pmatrix} 1 & 1 \\ 1 & 0 \end{pmatrix} \begin{pmatrix} 1 & 1 \\ 1 & 0 \end{pmatrix}$$
$$= \begin{pmatrix} 1 \times 1 + 1 \times 1 & 1 \times 1 + 1 \times 0 \\ 1 \times 1 + 0 \times 1 & 1 \times 1 + 0 \times 0 \end{pmatrix}$$
$$= \begin{pmatrix} 1+1 & 1+0 \\ 1+0 & 1+0 \end{pmatrix}$$
$$= \begin{pmatrix} 2 & 1 \\ 1 & 1 \end{pmatrix}$$

$$\begin{pmatrix} 1 & 1 \\ 1 & 0 \end{pmatrix}^3 = \begin{pmatrix} 2 & 1 \\ 1 & 1 \end{pmatrix} \begin{pmatrix} 1 & 1 \\ 1 & 0 \end{pmatrix}$$
$$= \begin{pmatrix} 2 \times 1 + 1 \times 1 & 2 \times 1 + 1 \times 0 \\ 1 \times 1 + 1 \times 1 & 1 \times 1 + 1 \times 0 \end{pmatrix}$$
$$= \begin{pmatrix} 2+1 & 2+0 \\ 1+1 & 1+0 \end{pmatrix}$$
$$= \begin{pmatrix} 3 & 2 \\ 2 & 1 \end{pmatrix}$$

$$\begin{pmatrix} 1 & 1 \\ 1 & 0 \end{pmatrix}^4 = \begin{pmatrix} 3 & 2 \\ 2 & 1 \end{pmatrix} \begin{pmatrix} 1 & 1 \\ 1 & 0 \end{pmatrix}$$
$$= \begin{pmatrix} 3 \times 1 + 2 \times 1 & 3 \times 1 + 2 \times 0 \\ 2 \times 1 + 1 \times 1 & 2 \times 1 + 1 \times 0 \end{pmatrix}$$
$$= \begin{pmatrix} 3+2 & 3+0 \\ 2+1 & 2+0 \end{pmatrix}$$
$$= \begin{pmatrix} 5 & 3 \\ 3 & 2 \end{pmatrix}$$

"Got it!" I said.

"Not yet! Not yet!" Tetra nearly shouted.

I looked up, and noticed that Lisa had changed seats. She was still staring at her computer, but at some point had moved to a chair right next to us.

"Want to work on this problem with us?" I asked.

"Done," she said, turning her computer toward me.

```
POWER(MATRIX(1,1,1,0),10) ⏎
⇒ MATRIX(89,55,55,34)
```

"Got it!" Tetra said. "The answer is $\left(\begin{smallmatrix} 89 & 55 \\ 55 & 34 \end{smallmatrix}\right)$!"

"Correct," Miruka said. "But did you find the hidden mystery?"

"Hidden mystery? Do you mean that—"

"Stop," Miruka said. She turned and pointed to Lisa. "How about you?"

Lisa grimly shook her head. Miruka twirled her finger like a conductor and pointed at Tetra.

"Back to you, then," she said.

"Right! We're creating a Fibonacci series! Like this."

$$1, \ 1, \ 2, \ 3, \ 5, \ 8, \ 13, \ \ldots$$

"I'll use F_n to represent the nth element in the series. Like, $F_1 = 1, F_2 = 1, F_3 = 2, F_4 = 3, F_5 = 5, \ldots$ and so on. Anyway, this matrix and the Fibonacci series are related, like this."

$$\begin{pmatrix} 1 & 1 \\ 1 & 0 \end{pmatrix}^n = \begin{pmatrix} F_{n+1} & F_n \\ F_n & F_{n-1} \end{pmatrix}$$

"We can easily prove this by mathematical induction," she continued. "The good part, where we show that if $n = k$ holds then

$n = k+1$ holds too, goes like this."

$$\begin{pmatrix} 1 & 1 \\ 1 & 0 \end{pmatrix}^{k+1} = \begin{pmatrix} 1 & 1 \\ 1 & 0 \end{pmatrix}^{k} \begin{pmatrix} 1 & 1 \\ 1 & 0 \end{pmatrix}$$

$$= \begin{pmatrix} F_{k+1} & F_k \\ F_k & F_{k-1} \end{pmatrix} \begin{pmatrix} 1 & 1 \\ 1 & 0 \end{pmatrix}$$

$$= \begin{pmatrix} F_{k+1} \times 1 + F_k \times 1 & F_{k+1} \times 1 + F_k \times 0 \\ F_k \times 1 + F_{k-1} \times 1 & F_k \times 1 + F_{k-1} \times 0 \end{pmatrix}$$

$$= \begin{pmatrix} F_{k+1} + F_k & F_{k+1} \\ F_k + F_{k-1} & F_k \end{pmatrix}$$

$$= \begin{pmatrix} F_{k+2} & F_{k+1} \\ F_{k+1} & F_k \end{pmatrix} \quad \text{use } F_{k+2} = F_{k+1} + F_k, F_{k+1} = F_k + F_{k-1}$$

"So calculating products of matrices is a lot like the recursion formula for the Fibonacci series."

$$\begin{cases} F_1 &= 1 \\ F_2 &= 1 \\ F_n &= F_{n-1} + F_{n-2} \quad (n \geqslant 3) \end{cases}$$

"Now we just need to find $\begin{pmatrix} 1 & 1 \\ 1 & 0 \end{pmatrix}^{10} = \begin{pmatrix} F_{11} & F_{10} \\ F_{10} & F_9 \end{pmatrix}$."

n	1	2	3	4	5	6	7	8	9	10	11	$\cdots$
F_n	1	1	2	3	5	8	13	21	34	55	89	$\cdots$

Answer 7-2 (exponentials of matrices)

$$\begin{pmatrix} 1 & 1 \\ 1 & 0 \end{pmatrix}^{10} = \begin{pmatrix} 89 & 55 \\ 55 & 34 \end{pmatrix}$$

"Well done, Tetra," Miruka said. "You found the Fibonacci series hidden within this problem, something that working the problem out by hand revealed. Lisa, on the other hand, used her computer to jump straight to the solution. That's not necessarily a bad thing, but it means she walked right over the buried treasure."

Lisa grimaced, but immediately recomposed herself. "It does."

Miruka smiled at Lisa's words, and pushed her glasses up her nose.

"There's more than one way of solving math problems. You can work them out in your head, do them by hand, use a computer... But regardless of how you do it, it's more fun when you can make discoveries on your way to the answer. It's more fun when you perceive hidden structures."

Miruka flipped her fingers 1, 1, 2, 3. Tetra and I shot her a 5 back.

"What's that?" Lisa asked.

"It's the Fibonacci sign!" Tetra said. "When somebody sends you a 1, 1, 2, 3, you send them a 5 back. We use a series that math fans love as a hand sign. You know, like a greeting."

Tetra flashed the sign to Lisa, who replied "Like this?" and raised her hand. Curiously, though, only her thumb and middle finger were extended, other fingers folded in.

"Uh, it's supposed to be a five..." Tetra said.

"00101 is five," Lisa said. "In binary."

7.4.2 Linear Transformations

"Let's talk about linear transformation," Miruka said. "About how we can consider products of matrices and vectors as representing movements of points." She opened my notebook to a new page and took the pencil from my hand. "Like you were talking about, the product of matrix $\begin{pmatrix} a & b \\ c & d \end{pmatrix}$ and vector $\begin{pmatrix} x \\ y \end{pmatrix}$ is this."

$$\begin{pmatrix} a & b \\ c & d \end{pmatrix} \begin{pmatrix} x \\ y \end{pmatrix} = \begin{pmatrix} ax + by \\ cx + dy \end{pmatrix}$$

"When we consider this to mean 'move point (x, y) to point $(ax + by, cx + dy)$,' that's called a linear transformation. Let's use the matrix $\begin{pmatrix} 2 & 1 \\ 1 & 2 \end{pmatrix}$ as an example."

$$\begin{pmatrix} 2 & 1 \\ 1 & 2 \end{pmatrix} \begin{pmatrix} x \\ y \end{pmatrix} = \begin{pmatrix} 2x + y \\ x + 2y \end{pmatrix}$$

"So here we're saying that matrix $\begin{pmatrix} 2 & 1 \\ 1 & 2 \end{pmatrix}$ moves point (x, y) to point $(2x + y, x + 2y)$. We're going to use the 'maps to' symbol to represent this movement, like this."

$$(x, y) \mapsto (2x + y, x + 2y)$$

"Here are some more examples."

$(0, 0) \mapsto (0, 0)$ because $\begin{pmatrix} 2 & 1 \\ 1 & 2 \end{pmatrix}\begin{pmatrix} 0 \\ 0 \end{pmatrix} = \begin{pmatrix} 0 \\ 0 \end{pmatrix}$

$(1, 0) \mapsto (2, 1)$ because $\begin{pmatrix} 2 & 1 \\ 1 & 2 \end{pmatrix}\begin{pmatrix} 1 \\ 0 \end{pmatrix} = \begin{pmatrix} 2 \\ 1 \end{pmatrix}$

$(0, 1) \mapsto (1, 2)$ because $\begin{pmatrix} 2 & 1 \\ 1 & 2 \end{pmatrix}\begin{pmatrix} 0 \\ 1 \end{pmatrix} = \begin{pmatrix} 1 \\ 2 \end{pmatrix}$

$(1, 1) \mapsto (3, 3)$ because $\begin{pmatrix} 2 & 1 \\ 1 & 2 \end{pmatrix}\begin{pmatrix} 1 \\ 1 \end{pmatrix} = \begin{pmatrix} 3 \\ 3 \end{pmatrix}$

$(-1, -1) \mapsto (-3, -3)$ because $\begin{pmatrix} 2 & 1 \\ 1 & 2 \end{pmatrix}\begin{pmatrix} -1 \\ -1 \end{pmatrix} = \begin{pmatrix} -3 \\ -3 \end{pmatrix}$

$(2, 1) \mapsto (5, 4)$ because $\begin{pmatrix} 2 & 1 \\ 1 & 2 \end{pmatrix}\begin{pmatrix} 2 \\ 1 \end{pmatrix} = \begin{pmatrix} 5 \\ 4 \end{pmatrix}$

$(100, 10) \mapsto (210, 120)$ because $\begin{pmatrix} 2 & 1 \\ 1 & 2 \end{pmatrix}\begin{pmatrix} 100 \\ 10 \end{pmatrix} = \begin{pmatrix} 210 \\ 120 \end{pmatrix}$

"Good so far?"

"Mmm, I think so?" Tetra said. "I'm just getting a little bit nervous about where this is heading."

"The Euclidian plane is nothing but a collection of points," Miruka continued. "The fact that we can use matrices to move a point means that we can also use them to transform the entire plane. Maybe Lisa can give us a little help."

Miruka leaned over to Lisa and whispered some instructions. Lisa nodded and started working on her computer. After a time, she turned it to show an array of dots on the monitor.

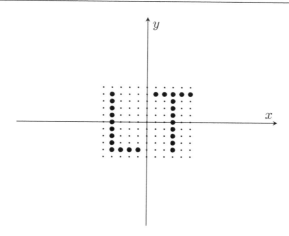

"Quickly done. I'm impressed. Lisa has drawn some dots so we can see how our linear transformation transforms the plane. And since dots alone can make it hard to see what's going on, we've added the 'LT' here."

"Standing for 'linear transformation,' I suppose?" Tetra said.

Miruka nodded and continued. "For example, the matrix $\begin{pmatrix} 2 & 1 \\ 1 & 2 \end{pmatrix}$ transforms the plane like this."

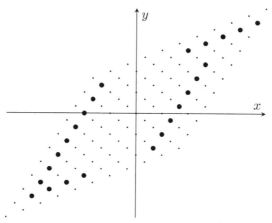

Linear transformation by matrix $\begin{pmatrix} 2 & 1 \\ 1 & 2 \end{pmatrix}$.

"Oh, look! It's all squished up!" Tetra said.

"It's doing what the matrix $\begin{pmatrix} 2 & 1 \\ 1 & 2 \end{pmatrix}$ tells it to do," Miruka said. "Different matrix, different transformation."

"So it's all up to the matrix. Makes perfect sense!"

"Okay, a quiz. Tetra, tell me how this matrix transforms the plane."

$$\begin{pmatrix} 1 & 0 \\ 0 & 1 \end{pmatrix}$$

"I need to think about where a point (x, y) moves to, right? Let's see..."

$$\begin{pmatrix} 1 & 0 \\ 0 & 1 \end{pmatrix} \begin{pmatrix} x \\ y \end{pmatrix} = \begin{pmatrix} x \\ y \end{pmatrix}$$

"So I guess the point just stays where it was?" Tetra said.

$$(x, y) \mapsto (x, y)$$

"Meaning?"

"Meaning the plane doesn't change at all?"

"Right. This is called the identity matrix, and it leaves the plane as it was. That's the identity transformation."

"Got it."

"Okay, next quiz. What does this matrix do to the plane?"

$$\begin{pmatrix} 0 & -1 \\ 1 & 0 \end{pmatrix}$$

"One minute..."

$$\begin{pmatrix} 0 & -1 \\ 1 & 0 \end{pmatrix} \begin{pmatrix} x \\ y \end{pmatrix} = \begin{pmatrix} -y \\ x \end{pmatrix}$$

"Huh, the x and the y get swapped, and the y goes negative."

$$(x, y) \mapsto (-y, x)$$

"So?"

"So I guess the plane gets flipped over?"

"Try it with a few points and see what happens," Miruka said.

"Okay, sure."

Tetra wrote in her notebook for a time, then looked up with a smile.

"Got it! It rotates the plane counterclockwise by 90°!"

"The center of rotation being...?"

"Uh, the origin!"

"Here," Lisa said.

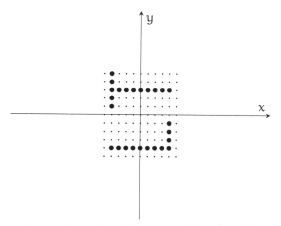

Linear transformation by matrix $\begin{pmatrix} 0 & -1 \\ 1 & 0 \end{pmatrix}$.

"I think I've noticed something, Miruka," Tetra said. "A linear transformation will never move the origin, right? Isn't that what this means?"

$$\begin{pmatrix} a & b \\ c & d \end{pmatrix} \begin{pmatrix} 0 \\ 0 \end{pmatrix} = \begin{pmatrix} 0 \\ 0 \end{pmatrix}$$

"In other words, $(0,0) \mapsto (0,0)$."

Miruka snapped her fingers. "A nice discovery," she said. "The origin is a fixed point. Put another way, you can't use a linear transformation to move the plane in an x–y direction, because that would move the origin."

"Got it."

"When you're looking at how a linear transformation behaves, points $(1,0)$ and $(0,1)$ are a good place to start," Miruka said. "You can use vectors $\begin{pmatrix} a \\ c \end{pmatrix}$ and $\begin{pmatrix} b \\ d \end{pmatrix}$ to represent where those points go. They directly show where the components of matrix $\begin{pmatrix} a & b \\ c & d \end{pmatrix}$ will move to, so looking at the matrix tells you what happens to points $(1,0)$ and $(0,1)$."

$$\begin{pmatrix} a & b \\ c & d \end{pmatrix} \begin{pmatrix} 1 \\ 0 \end{pmatrix} = \begin{pmatrix} a \\ c \end{pmatrix} \qquad \begin{pmatrix} a & \cdot \\ c & \cdot \end{pmatrix}$$

$$\begin{pmatrix} a & b \\ c & d \end{pmatrix} \begin{pmatrix} 0 \\ 1 \end{pmatrix} = \begin{pmatrix} b \\ d \end{pmatrix} \qquad \begin{pmatrix} \cdot & b \\ \cdot & d \end{pmatrix}$$

"Interesting," I said.

Miruka held up a finger. "Another quiz. After a linear transformation, will a plane always remain a plane?"

Tetra cocked her head. "Hmm, that's been the case so far. We've seen them stretched and spun around, but they're still planes." She stopped and thought some more, then said, "Actually, maybe that's not always what happens. For example..."

$$\begin{pmatrix} 0 & 0 \\ 0 & 0 \end{pmatrix}$$

"I think this matrix would squeeze everything into the origin, wouldn't it?"

"Indeed," Miruka said. "The zero matrix $\begin{pmatrix} 0 & 0 \\ 0 & 0 \end{pmatrix}$ will move any point to the origin."

I remembered the system of equations I had given to Yuri, my inconsistent and underdetermined examples, and wondered what would happen when using those.

"I think I might have another one," I said. "What about $\begin{pmatrix} 2 & 4 \\ 1 & 2 \end{pmatrix}$? That linear transformation won't result in a plane, will it?"

"Correct. That one will transform the plane into a straight line."

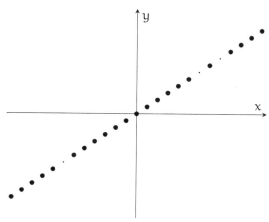

Linear transformation by matrix $\begin{pmatrix} 2 & 4 \\ 1 & 2 \end{pmatrix}$.

"It's all about the determinant, looks like," I said.

"That's right."

"How's that?" Tetra asked.

"If the determinant of the matrix describing the linear transformation is nonzero, then a plane will be transformed into a plane," I said. "If the determinant is zero, though, then points will be transformed to the origin, or to a line passing through the origin."

Plane $\mapsto$ Plane	Nonzero determinant
Plane $\mapsto$ Line through origin	Zero determinant (other than the zero matrix)
Plane $\mapsto$ Origin	Zero determinant (zero matrix)

"Hmm?" Tetra said. "The determinant was that thing that we used to find the solution to systems of equations, right?"

Miruka nodded. "You see multiplication of matrices and vectors when you're dealing with both systems of equations and $(x, y) \mapsto (s, t)$ linear transformations."

$$\begin{pmatrix} a & b \\ c & d \end{pmatrix} \begin{pmatrix} x \\ y \end{pmatrix} = \begin{pmatrix} s \\ t \end{pmatrix}$$

"Asking whether a system of equations $\begin{pmatrix} a & b \\ c & d \end{pmatrix}\begin{pmatrix} x \\ y \end{pmatrix} = \begin{pmatrix} s \\ t \end{pmatrix}$ has a unique solution and asking whether point $\begin{pmatrix} x \\ y \end{pmatrix}$ is the only point that a linear transformation moves to $\begin{pmatrix} s \\ t \end{pmatrix}$ is, in essence, the same thing. And in both cases, the answer is whether the determinant is nonzero."

7.4.3 Rotations

"Hey Miruka," I said. "About that matrix we looked at a minute ago, $\begin{pmatrix} 0 & -1 \\ 1 & 0 \end{pmatrix}$. That was a rotation matrix that results in a rotation by $\frac{\pi}{2}$ radians, yeah?"

$$\begin{pmatrix} \cos\frac{\pi}{2} & -\sin\frac{\pi}{2} \\ \sin\frac{\pi}{2} & \cos\frac{\pi}{2} \end{pmatrix}$$

"Of course," Miruka said.

"Whoa, hold up," Tetra said with some alarm. "What's with the trig functions all of a sudden?"

"What's the value of $\cos\frac{\pi}{2}$, Tetra?" Miruka asked.

"Um, let's see, $\frac{\pi}{2}$ is 90°, so... Right, that's 0."

Miruka raised an eyebrow. "Correct, but that took too long. You need to become better friends with radians and the trig functions. Okay, what's the value of $\sin\frac{\pi}{2}$?"

"Uh, uh, uh... 1!"

"Good. So you can understand this."

$$\begin{pmatrix} \cos\frac{\pi}{2} & -\sin\frac{\pi}{2} \\ \sin\frac{\pi}{2} & \cos\frac{\pi}{2} \end{pmatrix} = \begin{pmatrix} 0 & -1 \\ 1 & 0 \end{pmatrix}$$

"One second." Tetra pointed at each element in the matrix while reading them.

"The matrix $\begin{pmatrix} \cos\theta & -\sin\theta \\ \sin\theta & \cos\theta \end{pmatrix}$ is a counterclockwise rotation about the origin by θ radians," Miruka said. "So for $\theta = \frac{\pi}{2}$, the matrix $\begin{pmatrix} 0 & -1 \\ 1 & 0 \end{pmatrix}$ is a left rotation by $\frac{\pi}{2}$ radians around the origin, in other words 90° counterclockwise." She suddenly smiled. "And if $\theta = \frac{2\pi}{3}$, in other words if we have the matrix $\begin{pmatrix} \cos\frac{2\pi}{3} & -\sin\frac{2\pi}{3} \\ \sin\frac{2\pi}{3} & \cos\frac{2\pi}{3} \end{pmatrix}$..." Miruka trailed off and raised an eyebrow at me.

"It's—the Omega Waltz!" I said[1].

---

[1] See *Math Girls*, Chapter 3.

"The... what?" Tetra asked, looking back and forth between us.

"A dear old friend," Miruka said. "If you repeat a counterclockwise rotation by $\frac{2\pi}{3}$ radians, by 120° in other words, after three rotations you're back where you started from."

"Oh, yeah!" Tetra said. "Because $120° \times 3 = 360°$!"

"Which means the cube of a $\frac{2\pi}{3}$ radian rotation matrix is a complete rotation, making it equal to $\begin{pmatrix} 1 & 0 \\ 0 & 1 \end{pmatrix}$."

$$\begin{pmatrix} \cos\frac{2\pi}{3} & -\sin\frac{2\pi}{3} \\ \sin\frac{2\pi}{3} & \cos\frac{2\pi}{3} \end{pmatrix}^3 = \begin{pmatrix} 1 & 0 \\ 0 & 1 \end{pmatrix}$$

"Here," Lisa said, turning her computer around.

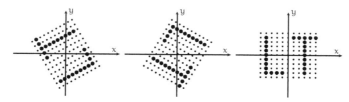

Rotations with $\theta = \frac{2\pi}{3}, \frac{4\pi}{3}, 2\pi$.

"This is the Omega Waltz, our triple-time dance," Miruka said.

"But this is theta, right? Where's the omega?" Tetra asked.

"Omega is a complex number that solves the third-degree equation $x^3 = 1$."

$$\omega = \frac{-1 + \sqrt{3}\,i}{2}$$

"So $\omega^3 = 1$," Miruka added.

"Oh, right."

"If we use the complex plane, we can create a correspondence between points and complex numbers."

Point (x, y)	←----→	Complex number $x + yi$
Rotation of $\frac{2\pi}{3}$ radians around the origin	←----→	Multiplication by ω
3 rotations of $\frac{2\pi}{3}$ radians around the origin	←----→	Multiplication by ω^3
$\begin{pmatrix} \cos\frac{2\pi}{3} & -\sin\frac{2\pi}{3} \\ \sin\frac{2\pi}{3} & \cos\frac{2\pi}{3} \end{pmatrix}^3 = \begin{pmatrix} 1 & 0 \\ 0 & 1 \end{pmatrix}$	←----→	$\omega^3 = 1$

"Rotations, huh," I said. I could practically feel matrices, linear transformations, and complex numbers making new connections within my head. "Points on the plane rotating around the origin kind of reminds me of stars rotating about the North Star. Stars like points in a plane, constellations like shapes in the plane, the plane itself forming the sky..."

"Like a planetarium!" Tetra said.

"With rotation matrices spinning an infinite plane around and around," Miruka said, giving her pen a twirl around her finger. "A nice image."

7.5 Heading Home

7.5.1 Dialogues

The four of us—Miruka, Tetra, Lisa, and myself—decided to walk to the station together.

"So matrices are an arrangement of numbers," Tetra said. "We can use them to represent systems of equations, or to represent linear transformations. Kinda makes me wonder which one they *really* are."

"They really are both," Miruka replied. "Actually, since you're representing systems of equations and matrices in the form of matrices, it's the matrix itself that's the true form. In other words, all things that you can represent as a matrix have some common characteristics. As you study math, you'll find plenty more things that you can represent as a matrix. Once you've learned to identify them, the logic of matrices can be quite a powerful tool."

"A new addition to my toolbox!" Tetra said.

"And another one to keep in good shape. Without care and use, even the best tools will become dull and rusty. In math, that means memorization isn't enough."

I recalled a line from a poem by Hideo Kobayashi—

> It is not enough to memorize.
>
> One must also remember.

"We pose problems, and solve them," Miruka said, in nearly a chant. "We find mysteries, and solve those. It is through these dialogues that we sharpen our mathematical tools."

Dialogues, huh. I can see that. Becoming engaged in a problem is like having a dialogue with it. And Miruka, Tetra, and I solve problems through dialogue. Those dialogues allow me to measure my understanding, and know my own abilities. Even our slogan, "Examples are key to understanding," is an invitation to dialogue, a call for creating an example showing that we understand what we're doing...

"Interesting, that's how I feel when I'm reading," Tetra said, snapping me out of my reverie. "Like I'm in a dialogue with the author. In that sense, I guess learning is something like an accumulation of dialogues."

"There are two kinds of solitude," Miruka said. "Solitude with dialogue, and solitude without dialogue."

Solitude with dialogue? I thought. Solitude without dialogue I could understand, but...

I wondered if when Miruka's eyes were closed she was in a dialogue with herself, or perhaps with her memories.

"So long as there is dialogue, even solitude is not idleness," she said.

"When I look at matrices," Tetra said, fists clenched, "I can see how so many things are connected—matrices, points, lines, planes... It also feels like there's a lot more to determinants that I haven't seen yet. Maybe I'll pick up a book about this stuff, so you guys don't have to teach me *everything* about it. In fact, it seems like I saw a book about linear algebra just recently, but where... Ah—"

Tetra stole a glance at me, then looked down at her feet, her face turning red.

The various components of linear algebra—vector spaces, matrices, linear mappings, simultaneous equations, even linear and planar equations—all stand upon the stage of "linearity."

KOJI SHIGA [15]

My notes

The linearity of linear transformations

The following shows the linearity of a linear transformation by a 2×2 matrix.

The linear transformation of a sum is the sum of the linear transformations

The result of using matrix $\begin{pmatrix} a & b \\ c & d \end{pmatrix}$ to perform a linear transformation on a sum of two vectors $\begin{pmatrix} s \\ t \end{pmatrix}$ and $\begin{pmatrix} v \\ w \end{pmatrix}$ is the same as performing the linear transformation on the vectors independently and adding the results.

$$\begin{pmatrix} a & b \\ c & d \end{pmatrix} \left(\begin{pmatrix} s \\ t \end{pmatrix} + \begin{pmatrix} v \\ w \end{pmatrix} \right) = \begin{pmatrix} a & b \\ c & d \end{pmatrix} \begin{pmatrix} s+v \\ t+w \end{pmatrix}$$

$$= \begin{pmatrix} a(s+v) + b(t+w) \\ c(s+v) + d(t+w) \end{pmatrix}$$

$$= \begin{pmatrix} (as+bt) + (av+bw) \\ (cs+dt) + (cv+dw) \end{pmatrix}$$

$$= \begin{pmatrix} a & b \\ c & d \end{pmatrix} \begin{pmatrix} s \\ t \end{pmatrix} + \begin{pmatrix} a & b \\ c & d \end{pmatrix} \begin{pmatrix} v \\ w \end{pmatrix}$$

We therefore have the following:

$$\underbrace{\begin{pmatrix} a & b \\ c & d \end{pmatrix} \underbrace{\left(\begin{pmatrix} s \\ t \end{pmatrix} + \begin{pmatrix} v \\ w \end{pmatrix} \right)}_{\text{sum}}}_{\text{linear transformation}} = \underbrace{\underbrace{\begin{pmatrix} a & b \\ c & d \end{pmatrix} \begin{pmatrix} s \\ t \end{pmatrix}}_{\text{linear transformation}} + \underbrace{\begin{pmatrix} a & b \\ c & d \end{pmatrix} \begin{pmatrix} v \\ w \end{pmatrix}}_{\text{linear transformation}}}_{\text{sum}}$$

The linear transformation of a scalar multiple is the scalar multiple of the linear transformation

The result of using matrix $\begin{pmatrix} a & b \\ c & d \end{pmatrix}$ to perform a linear transformation on a vector $\begin{pmatrix} s \\ t \end{pmatrix}$ multiplied by a scalar value K is the same as performing the linear transformation on the vector and multiplying the result by K.

$$\begin{pmatrix} a & b \\ c & d \end{pmatrix} \left(K \begin{pmatrix} s \\ t \end{pmatrix} \right) = \begin{pmatrix} a & b \\ c & d \end{pmatrix} \begin{pmatrix} Ks \\ Kt \end{pmatrix}$$

$$= \begin{pmatrix} aKs + bKt \\ cKs + dKt \end{pmatrix}$$

$$= K \begin{pmatrix} as + bt \\ cs + dt \end{pmatrix}$$

$$= K \begin{pmatrix} a & b \\ c & d \end{pmatrix} \begin{pmatrix} s \\ t \end{pmatrix}$$

We therefore have the following:

$$\underbrace{\begin{pmatrix} a & b \\ c & d \end{pmatrix} \underbrace{\left(K \begin{pmatrix} s \\ t \end{pmatrix} \right)}_{\text{scalar mult.}}}_{\text{linear transformation}} = K \underbrace{\underbrace{\begin{pmatrix} a & b \\ c & d \end{pmatrix} \begin{pmatrix} s \\ t \end{pmatrix}}_{\text{linear transformation}}}_{\text{scalar mult.}}$$

The above furthermore holds for an $n \times n$ matrix.

CHAPTER **8**

A Random Walk Alone

> I got over the fence, and laid me down in the shade to rest my limbs, for I was very weary, and fell asleep; but judge you, if you can, that read my story, what a surprise I must be in when I was awaked out of my sleep by a voice calling me by my name several times, "Robin, Robin, Robin Crusoe: poor Robin Crusoe! Where are you, Robin Crusoe? Where are you? Where have you been?"
>
> DANIEL DEFOE
> *Robinson Crusoe*

8.1 AT HOME

8.1.1 A Rainy Saturday

The next Saturday it was raining, as it had been for several days. Even worse it had gotten hot, making the humidity awful. I started my day by trying to study in my room, as usual, and as usual I failed. Yuri was there, making her presence impossible to ignore.

"What's up with you today? I asked.

"Nothing," she sighed.

Since starting her third year in junior high school, Yuri had started bringing textbooks and homework with her. My room had become her weekend study hall. She had brought some notebooks with her that day too, but didn't seem to be making much progress in her studies. I tried probing into what was bothering her a couple of times, but only got vague, uninformative answers.

She finally let loose another huge sigh, and said, "I hate the rain."

8.1.2 Teatime

My mother appeared bearing snacks: jellies made from sweet azuki beans.

"How's the homework going, Yuri?" she asked.

Yuri suddenly perked up, as she often did when speaking to my mother. "Great! I've got him here to help me when I need it," Yuri said, pointing a thumb at me. I decided not to mention that I hadn't seen her study at all, much less ask me questions.

"Be nice when you're helping her," my mother said to me.

"Always am," I replied.

"These are delicious!" Yuri said.

"I'm glad. This is the perfect season for them."

"Speaking of which, your hydrangea arrangement in the hallway is just gorgeous!"

"I'm so glad you noticed!" mother said, beaming.

Really pouring it on kinda thick there, I thought.

8.1.3 Piano Problem

Rejuvenated by the snack, Yuri pulled a book of mathematical brainteasers off my bookshelf and flipped through the pages.

"Here's a good one for you," she said.

Problem 8-1 (The piano problem)

Connect notes from adjacent white keys on a piano keyboard to create a melody, under the following conditions:

- Start with C and end three notes higher, on F.
- A melody comprises 12 notes.
- No notes lower than the starting note can be used.

Example: C → D → C → D → E → D → C → D → E → F → G → F is a valid melody. However, C → D → E → C → D → E → E → ... is invalid, because it includes steps that do not involve adjacent keys (E → C and E → E).

How many melodies can be created?

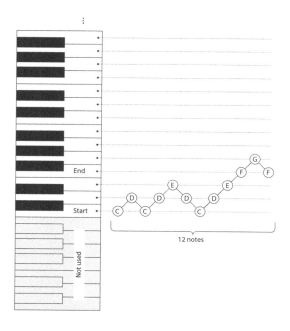

"That's... actually kind of interesting," I said.

"Toldja." Yuri grinned. "We have to use permutations and combinations, right?"

"Probably. That's jumping the gun, though. The first step is to make sure we really understand the problem. Starting on a problem that you've misunderstood can only waste your time."

"Well, duh. Obviously you can't solve a problem if you don't know what it is."

"Indeed. You'd be surprised how easy it is to fall into that trap, though."

"Okay, so what do we do for this piano problem?"

I pulled out a sheet of graph paper.

"As usual, examples are the key to understanding," I said. "The problem wants to know how many melodies we can create, subject to some conditions. So we should create a melody or two to make sure we understand those conditions. We might also stumble across some clues as to how to solve the problem along the way."

"Sounds like a plan!" Yuri said. She put on her glasses, and peered over my shoulder at the graph paper.

8.1.4 Example Melody

"The problem has already given us one example, here," I said.

$$C \to D \to C \to D \to E \to D \to C \to D \to E \to F \to G \to F$$

"Let's start by graphing that."

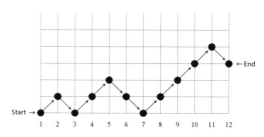

An example melody (from the problem statement).

"It's a zig-zag," Yuri said.

"Can you think of any other examples?"

"Easy peasy. You could go C → D → E → F and so on to get as high as possible, then come back down like → B → A → G → F at the end."

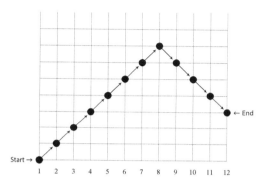

An example melody that goes as high as possible, then back down.

"Looks good," I said. "You start at C and end on F, and there's 12 notes, so you've met the conditions. Can you think of another?"

"I guess I could do the opposite—go low and then high. Oops, or not. Can't go below C. Well then, how about starting out C → D → C → D → C, flying under the radar, then making a sudden ascent."

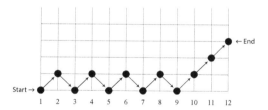

An example melody that starts out low.

"An excellent example," I said.

"I noticed something when drawing that graph," Yuri said. "Any melody that meets the conditions will have to go up 7 times, and down 4 times. See? It's obvious if you color the arrows!"

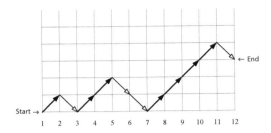

"That's actually a pretty amazing thing to notice," I said.

"Oh, I get it!" Yuri said, giving her ponytail a shake. "The number of melodies we can create will be the number of permutations of 7 up and 4 down arrows!"

"That's a bold assertion," I said. "I wonder if it's really true, though."

"How couldn't it be? I mean... Oh, wait. I still can't go below the first note, C. If we just calculated the permutations, we would be counting ones that go too low, too."

"Exactly," I said. "You can fly under the radar, but not underground."

"Well you seem pretty sure of yourself. Do you know what the answer is?"

"I'm not sure what the exact answer is yet, but I've found two ways to solve the problem."

"*Two* ways?"

8.1.5 Solution 1: Hard Work

"First, let's solve it using a Tetra approach," I said.

"A Tetra approach? What's that?" Yuri asked.

"That's where you just hammer away at the problem until you arrive at the solution. In this case, we start at the leftmost note, and for each step in turn we count the number of ways to get to that note."

I pulled out a new sheet of graph paper and started writing some numbers.

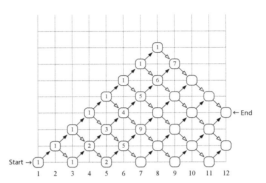

Yuri squinted at the page. "And just what sophisticated technique is this?"

"It's called adding," I said.

"Oh... Oh! You're just adding up the number of ways to get to each note from above and from below!"

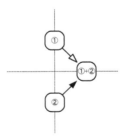

"Okay, I got this. Scoot," Yuri said. She took the pencil from my hand and finished filling in the numbers.

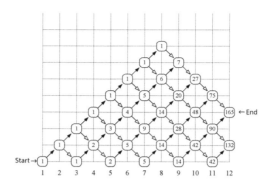

"Perfect," I said. "Well done. So the answer is...?"

"165! There are 165 ways to end up on F as the twelfth note. In other words, we can make 165 melodies under these conditions."

Answer 8-1 (The piano problem)

There are 165 possible melodies.

8.1.6 Solution 2: Smart Work

"So the other way to solve this problem is—" I began.

"The Miruka approach, right?" Yuri said.

"Er, maybe. Anyway, start out by going as low as you can."

"But then you go below where you started from! That's against the rules!"

"Yeah, but listen through to the end. Let's let P be the starting note, and Q be the end note. Any path that goes below the starting note on its way from P to Q will definitely pass through at least one of these four points, R1, R2, R3, R4."

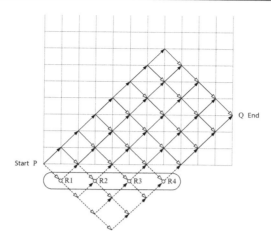

"Huh? Hang on a sec," Yuri said. She took a few seconds to stare at the graph, then nodded. "Okay, got it. And?"

"Next, we want to create a horizontal mirror that passes through these four points, and think about point Q', the reflection of our end point Q."

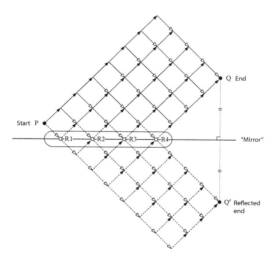

"Whoa now," Yuri said. "Just what are we supposed to learn from this?"

"You tell me."

Yuri's ponytail gleamed golden in the sun as she thought. I backed off and gave her space to work. While waiting, I noticed how much she had grown in her final year of junior high, and how her expressions were becoming more adult.

"I give up," she said. "I surrender. Tell me, what am I supposed to see here?"

"Okay, so we're trying to find the number of ways to go from the first point P to the final point Q, right?"

"Except that we can't count all the paths that go below where we started from."

"Right. Which means we just need to count the number of those illegal paths and subtract them out."

"Leading to the obvious question, just how do we do that?"

"Look at the figure. Every path from P to Q that goes below the first note passes through at least one of these four points, R1, R2, R3, R4."

"We've been through this."

"Right. But think about the mirror. The number of paths through the R points from P to Q will be the same as the number of paths from P to Q'."

"Where on earth do you get that from?"

"Look. If we're passing through the mirror on our way from P to Q, then after hitting one of R1, R2, R3, R4, our up–down direction of travel will flip. In other words, an up–right movement will become a down–right movement, and a down–right movement becomes up–right. Just like a reflection in a mirror. So there's a one-to-one correspondence between paths from P to Q' and paths starting at P and going below the first note to get to Q."

"Oh ho!"

"In other words, the number of paths from P to Q' is the number of paths we shouldn't consider, the one we should subtract to get our answer."

I quickly sketched out an overview of the number of paths.

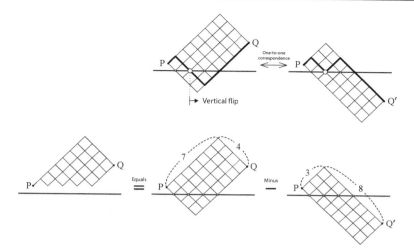

"Very cool!" Yuri said. Then her eyes narrowed, and she gave me a suspicious look. "Actually, a little bit too cool. Where'd you get this idea?"

"Uh, from Miruka."

"I knew it!"

"Anyway, the total number of paths from P to Q is equal to the number of ways we can order 7 up arrows and 4 down arrows. That would be the number of ways of selecting 7 out of 11 arrows to be designated as 'up' arrows. In school we write that $_{11}C_7$, but here let's write it as $\binom{11}{7}$."

Number of paths from P to Q
= Number of ways of selecting 7 of $7 + 4$ arrows as 'up' arrows
$=_{7+4}C_7$
$= \binom{7+4}{7}$
$= \binom{11}{7}$
$= \binom{11}{4}$ selecting 7 of 11 is the same as selecting 4 of 11
$= \dfrac{11 \cdot 10 \cdot \cancel{9}^{3} \cdot \cancel{8}}{\cancel{4} \cdot \cancel{3} \cdot \cancel{2} \cdot 1}$
$= 11 \cdot 10 \cdot 3$
$= 330$

"So in all there are 330 paths from P to Q," I said.

"Wow, way more than I would have thought!"

"Next we need the number of paths from P to Q'. This time we're looking for the number of ways to arrange 3 up arrows and 8 down arrows."

Number of paths from P to Q'
= Number of ways of selecting 3 of $3 + 8$ arrows as 'up' arrows
$= \binom{3+8}{3}$
$= \binom{11}{3}$
$= \dfrac{11 \cdot \cancel{10}^{5} \cdot \cancel{9}^{3}}{\cancel{3} \cdot \cancel{2} \cdot 1}$
$= 11 \cdot 5 \cdot 3$
$= 165$

"So there are 165 paths from P to Q'," I said.

"And now we just subtract."
"Right."

Number of paths from P to Q not lower than initial note
= Paths from P to Q − Paths from P to Q′
$$= \binom{7+4}{4} - \binom{3+8}{3}$$
$$= 330 - 165$$
$$= 165$$

"So the answer to the piano problem is 165," I said. "The same answer we got when we did it the 'hard' way."

"An exact match! Wow, that feels good."

8.1.7 Generalization

"Actually, we can go a step further and generalize the number of melodies we can create," I said.

"Meaning what?" Yuri said.

"Meaning we can find a way to calculate the number of ways of making melodies of any length. That would be difficult to do using the 'hard' method, but with the 'smart' method we can create a formula that will let us easily calculate the number of melodies."

"And it works in pretty much the same way as what we just did?"

"Sure, except that we want to use variables in place of specific values. In the original problem, for example, we wanted to go 7 notes up and 4 notes down, but instead let's consider the case where we want to go u notes up and d notes down. This is called 'generalization through introduction of variables.' We'll do it like this."

Number of paths from start to end not lower than initial note
= Paths from start to end − Paths from start to reflected end

$$
\begin{aligned}
&= \binom{u+d}{d} - \binom{(d-1)+(u+1)}{d-1} \\
&= \binom{u+d}{d} - \binom{u+d}{d-1} \\
&= \frac{(u+d)!}{d!\,(u+d-d)!} - \frac{(u+d)!}{(d-1)!\,(u+d-(d-1))!} \\
&= \frac{(u+d)!}{u!\,d!} - \frac{(u+d)!}{(u+1)!\,(d-1)!}
\end{aligned}
$$

"See? It's just like that last step we did for the problem," I said.

"Except way messier and pretty much impossible to calculate. What am I supposed to do with this?"

$$\frac{(u+d)!}{u!\,d!} - \frac{(u+d)!}{(u+1)!\,(d-1)!} = ?$$

"By finding a common denominator, just like you always do when subtracting fractions."

"A common denominator? Of *this*?"

"It's not that bad. For one thing, we can pull a $(d-1)!$ out of $d!$, which makes things easier."

$$d! = d \cdot \underbrace{(d-1) \cdot (d-2) \cdot \ldots \cdot 2 \cdot 1} = d \cdot \underbrace{(d-1)!}$$

"Let me see... Okay, sure, I see that. Makes sense that d times $(d-1)!$ equals d!, I suppose."

"Similarly, $u+1$ times $u!$ equals $(u+1)!$, right? So let's use that to create a common denominator."

$$\frac{(u+d)!}{u!\,d!} - \frac{(u+d)!}{(u+1)!\,(d-1)!}$$
$$= \frac{u+1}{u+1} \cdot \frac{(u+d)!}{u!\,d!} - \frac{d}{d} \cdot \frac{(u+d)!}{(u+1)!\,(d-1)!}$$
$$= \frac{(u+1)(u+d)!}{\underwave{(u+1)u!}\,d!} - \frac{d(u+d)!}{(u+1)!\,\underwave{d(d-1)!}}$$
$$= \frac{(u+1)(u+d)!}{\underwave{(u+1)!}\,d!} - \frac{d(u+d)!}{(u+1)!\,\underwave{d!}}$$
$$= \frac{(u+1) \cdot (u+d)! - d \cdot (u+d)!}{(u+1)!\,d!}$$

"Okay, what do you think we should do next?" I asked.

"Calculate out the numerator, I guess? Oh, I see! We can factor out a $(u+d)!$, can't we?"

$$\frac{(u+1) \cdot (u+d)! - d \cdot (u+d)!}{(u+1)!\,d!} = \frac{((u+1) - d)(u+d)!}{(u+1)!\,d!}$$
$$= \frac{(u-d+1)(u+d)!}{(u+1)!\,d!}$$

"Hmm..."

"'Hmm' what? We're done. It's generalized."

"Yeah, but something tells me we can clean this up even further."

$$\frac{(u-d+1)(u+d)!}{(u+1)!\,d!}$$

"Gah! You never quit. You're like a mathematical neat freak."

"Quiet for a minute, I'm thinking."

I focused on the equation, recalling how I had recently told Tetra not to let complexity get in her way. I was determined to follow my own advice here.

"Well?" Yuri said.

"Well, we've got factorials in both the numerator and the denominator. I think if we multiply both by $u + d + 1$, we could do something with combinations."

$$\frac{(u-d+1)(u+d)!}{(u+1)!\,d!} = \frac{u+d+1}{u+d+1} \cdot \frac{(u-d+1)(u+d)!}{(u+1)!\,d!}$$

$$= \frac{(u-d+1)(u+d+1)(u+d)!}{(u+d+1)(u+1)!\,d!}$$

$$= \frac{(u-d+1)(u+d+1)!}{(u+d+1)(u+1)!\,d!}$$

$$= \frac{u-d+1}{u+d+1} \cdot \frac{(u+d+1)!}{(u+1)!\,d!}$$

$$= \frac{u-d+1}{u+d+1} \cdot \frac{(u+d+1)!}{(u+1)!\,(u+d+1-(u+1))!}$$

$$= \frac{u-d+1}{u+d+1} \cdot \binom{u+d+1}{u+1}$$

"Okay," Yuri said. "I guess that's an improvement? Maybe?"

$$\frac{u-d+1}{u+d+1} \cdot \binom{u+d+1}{u+1}$$

I still wasn't satisfied, though, so I kept staring at what I had. I knew one good way of simplifying complex statements was looking for common patterns, and pretty soon I had found one.

"How about we make a substitution like this?" I said.

$$\begin{cases} a = u+1 \\ b = d \end{cases}$$

"Which does what?" Yuri asked.

$$\frac{u-d+1}{u+d+1} \cdot \binom{u+d+1}{u+1} = \frac{(u+1)-d}{(u+1)+d} \cdot \binom{(u+1)+d}{u+1}$$

$$= \frac{a-b}{a+b} \cdot \binom{a+b}{a}$$

"Okay, now I'm happy," I said.

$$\frac{a-b}{a+b} \cdot \binom{a+b}{a}$$

"Well done!" Yuri clapped her hands. "Bravo!"

"Why thank you," I said. "I'm glad someone was here to appreciate it."

I paused to write this up in my notes.

> **Generalization of the piano problem**
>
> The number of melodies that can be created by starting from one note and connecting $a+b$ adjacent notes, never going lower than the starting note and ending at a note $a-b-1$ notes higher than the starting note, is
> $$\frac{a-b}{a+b} \cdot \binom{a+b}{a}.$$

"Ending at a note $a - b - 1$ notes higher? Huh?" Yuri said.

"The final note is $u - d$ notes higher that what we started at. In other words, $u - d = (a-1) - b = a - b - 1$ notes higher. There are $u + d + 1$ notes in all, which is $u + d + 1 = (a-1) + b + 1 = a + b$."

"If you say so."

"Now that we have a generalization, let's try putting into use. The problem in the book used $a - b - 1 = 3$ and $a + b = 12$, so $a = 8$ and $b = 4$. So..."

$$\begin{aligned}
\frac{a-b}{a+b} \cdot \binom{a+b}{a} &= \frac{8-4}{8+4} \cdot \binom{8+4}{8} \\
&= \frac{4}{12} \cdot \binom{12}{8} \\
&= \frac{4}{12} \cdot \binom{12}{4} \\
&= \frac{\cancel{4}}{\cancel{12}} \cdot \frac{\cancel{12} \cdot 11 \cdot \overset{5}{\cancel{10}} \cdot \overset{3}{\cancel{9}}}{\cancel{4} \cdot \cancel{3} \cdot \cancel{2} \cdot 1} \\
&= 11 \cdot 5 \cdot 3 \\
&= 165
\end{aligned}$$

"Whoa!" Yuri shouted. "It worked! 165 on the nose!"

8.1.8 Apprehension (1)

"Well that was more fun than expected," I said as I straightened up my desk.

"I must agree," Yuri said. Her expression was much brighter than before.

"You feeling better?" I asked.

"Well, I was until you reminded me," she said, sighing. "Thanks, cuz." She took off her glasses and started playing with them. "I guess sending that letter was a mistake after all."

So that's what's going on. Yuri's friend—'that guy,' the one who had moved. I guess she followed Tetra's advice and sent a letter...

Yuri was spinning her glasses in circles on my desk.

"I guess you didn't get an answer?"

"Nah." Yuri stood and walked to my bookcase to browse. "It's no big deal," she said, her back to me.

Obviously it was. I'd never really been in her situation, but I could imagine how uncomfortable it must be—waiting for an answer that might never come.

I spoke in the kindest way I could. "He's only just moved, right? I'm sure he's just been too busy, new school and all."

"Huh. Yeah, maybe you're right." Yuri turned and smiled. "Thank you."

8.2 On the Way to School

8.2.1 Random Walks

"Good morning!"

"Hey, Tetra."

On Monday morning it was still raining. Tetra and I had run into each other at the train station, so we started walking to school together. Tetra was carrying a bright orange umbrella.

"How can you be so bright and chipper even in the rain?" I asked.

"Appearances can be deceiving, you know."

"What, is something bothering you?"

"No, I'll be fine, once I turn Miruka down."

"I must be missing something here..."

"It's nothing, don't worry. What did you do over the weekend?"

I gave Tetra a brief rundown of the piano problem I had worked out with Yuri.

"Making zig-zag paths, huh?" she said, sidestepping to avoid a puddle. "Sounds a lot like a random walk. You know, where you randomly move from point to point?"

"Yeah, kinda."

"We just recently studied those in my physics class, when we were talking about Brownian motion. We watched this neat video where pollen absorbed enough water to burst, then scattered this fine powder that danced about. Talking about moving up and down on a keyboard reminded me of that."

"Sure, I can see that," I said, impressed with how Tetra's imagination could make such connections. "I guess you could even make a one-dimensional random walk on a keyboard, flipping a coin at each key and moving higher or lower depending on the result."

"A one-dimensional keyboard?"

"The surface of the keyboard itself is two-dimensional, if you're thinking of it as having up–down and right–left directions, but the sounds it makes can only go high–low, so the numeric model for the random walk would be one-dimensional."

"I see," Tetra said, nodding several times.

8.3 Lunch in Our Classroom

8.3.1 Practicing with Matrices

After our morning classes ended, it was time for lunch. Tetra often brought her bento box to our classroom to eat, an uncharacteristically bold move—she was in her second year while Miruka and I were in our third, and most lower-level students would be hesitant to step into the realm of their elders. It didn't seem to bother her, though.

"Where's Miruka?" she asked.

"Absent today," I said.

"Oh, I see."

"Want to eat in here anyway? It's raining outside, so..."

Tetra smiled and sat in an empty chair next to me.

"I've been studying matrices more," she said as she opened her lunchbox. "Since you taught me that mnemonic I never get rows and columns confused any more, so thanks for that! I'm also getting better at finding patterns in equations, especially sums of products. It's kinda fun, actually!"

"Well let's see how you're doing, then. Can you calculate this?"

$$\begin{pmatrix} a & b \\ c & d \end{pmatrix}^2$$

"Oh, come on," Tetra said. "That's too easy."

"Yeah?"

$$\begin{pmatrix} a & b \\ c & d \end{pmatrix}^2 = \begin{pmatrix} a & b \\ c & d \end{pmatrix} \begin{pmatrix} a & b \\ c & d \end{pmatrix}$$
$$= \begin{pmatrix} aa + bc & ab + bd \\ ca + dc & cb + dd \end{pmatrix}$$
$$= \begin{pmatrix} a^2 + bc & (a+d)b \\ (a+d)c & cb + d^2 \end{pmatrix}$$

"Okay, so that wasn't so hard. By the way, see how these bits with $a + d$ pop out?"

"Sure, the $(a+d)b$ and the $(a+d)c$ here."

"Okay, so another calculation. Try this one."

$$(a+d) \begin{pmatrix} a & b \\ c & d \end{pmatrix}$$

"Can't fool me," Tetra said. "This is just multiplication of a matrix by some number. So I just have to multiply all the elements in the matrix by $a + d$.

$$(a+d) \begin{pmatrix} a & b \\ c & d \end{pmatrix} = \begin{pmatrix} (a+d)a & (a+d)b \\ (a+d)c & (a+d)d \end{pmatrix}$$

"Very good," I said. "Not that I was trying to trick you or anything, but it's interesting to look at these together."

$$\begin{cases} \begin{pmatrix} a & b \\ c & d \end{pmatrix}^2 = \begin{pmatrix} a^2 + bc & (a+d)b \\ (a+d)c & cb + d^2 \end{pmatrix} \\ (a+d) \begin{pmatrix} a & b \\ c & d \end{pmatrix} = \begin{pmatrix} (a+d)a & (a+d)b \\ (a+d)c & (a+d)d \end{pmatrix} \end{cases}$$

Tetra did just as I suggested, and stared at what I had written in my notebook.

"Well, I see that they both have $(a+d)b$ and $(a+d)c$ components, but..."

"Actually that's exactly what I hoped you would notice. Now give this a try."

$$\begin{pmatrix} a & b \\ c & d \end{pmatrix}^2 - (a+d)\begin{pmatrix} a & b \\ c & d \end{pmatrix}$$

"Okay, give me a minute," she said, and started writing.

$$\begin{pmatrix} a & b \\ c & d \end{pmatrix}^2 - (a+d)\begin{pmatrix} a & b \\ c & d \end{pmatrix} = \begin{pmatrix} a^2+bc & (a+d)b \\ (a+d)c & cb+d^2 \end{pmatrix} - \begin{pmatrix} (a+d)a & (a+d)b \\ (a+d)c & (a+d)d \end{pmatrix}$$

$$= \begin{pmatrix} a^2+bc-(a+d)a & (a+d)b-(a+d)b \\ (a+d)c-(a+d)c & cb+d^2-(a+d)d \end{pmatrix}$$

$$= \begin{pmatrix} a^2+bc-a^2-da & 0 \\ 0 & cb+d^2-ad-d^2 \end{pmatrix}$$

$$= \begin{pmatrix} bc-da & 0 \\ 0 & cb-ad \end{pmatrix}$$

"Oh, neat!" Tetra said. "Two of the elements get knocked out when we subtract! Well, maybe I shouldn't be so surprised, since we're doing subtraction and all."

"You're right that two elements have disappeared, but what else happened?"

"Um, the $bc - da$ and the $cb - ad$ don't disappear?"

"They don't disappear and...?" I waited to see if she would notice what she was missing.

"And, uh..."

"Look carefully at the $bc - da$ and the $cb - ad$."

"The $bc - da$ and the $cb - ad$... Oh! Oh! They're equal! I've just flipped the order of the letters!"

"They're equal and...?"

"What, there's more?" Tetra blinked rapidly and looked at me.

"Think about how $bc - da$ equals $-(ad - bc)$," I said.

"And $-(ad - bc)$ is... Oh! $ad - bc$ is the determinant!"

"Exactly. So if we let $\left|\begin{smallmatrix} a & b \\ c & d \end{smallmatrix}\right|$ be the determinant of $\left(\begin{smallmatrix} a & b \\ c & d \end{smallmatrix}\right)$, then we get this."

$$\begin{pmatrix} a & b \\ c & d \end{pmatrix}^2 - (a+d)\begin{pmatrix} a & b \\ c & d \end{pmatrix} = -\begin{vmatrix} a & b \\ c & d \end{vmatrix}\begin{pmatrix} 1 & 0 \\ 0 & 1 \end{pmatrix}$$

"Huh," Tetra said.

"Let's move everything to the left and see what happens."

$$\begin{pmatrix} a & b \\ c & d \end{pmatrix}^2 - (a+d)\begin{pmatrix} a & b \\ c & d \end{pmatrix} + \begin{vmatrix} a & b \\ c & d \end{vmatrix}\begin{pmatrix} 1 & 0 \\ 0 & 1 \end{pmatrix} = \begin{pmatrix} 0 & 0 \\ 0 & 0 \end{pmatrix}$$

"Wow, that's kind of pretty," Tetra said.

"Now let's let $\left(\begin{smallmatrix} a & b \\ c & d \end{smallmatrix}\right)$ be A, let $\left(\begin{smallmatrix} 1 & 0 \\ 0 & 1 \end{smallmatrix}\right)$ be E, and let $\left(\begin{smallmatrix} 0 & 0 \\ 0 & 0 \end{smallmatrix}\right)$ be O. Then we get this."

$$A^2 - (a+d)A + (ad - bc)E = O$$

"This equation will always hold. It even has a name, the Cayley–Hamilton theorem. It's a good one to remember for college entrance exams. Questions based on this theorem are pretty common."

"I'll be sure to learn it!" Tetra said. "By the way, we call $ad - bc$ the determinant, but does $a + d$ have a name too?"

"Huh, I'd never thought of that. Not that I know of..."

Miruka's absence made itself all the more noticeable. I was sure that if she were there, and if $a + d$ did have a name, she'd be able to tell us right off.

8.3.2 Apprehension (2)

We set matrices aside to focus on our lunches. I unwrapped a roll as Tetra unpacked her bento.

"By the way," I said, "what were you talking about this morning, something about turning Miruka down?"

"Oh, that," Tetra said, seemingly unsure what she should say. "Just something Miruka had asked me to do. She wanted me to substitute for her, something about a conference presentation."

"A conference? Seriously?"

"Just a small one, this summer at the Narabikura Library. They're hosting an international computer science conference."

"Wow, and you're going to present a paper there?"

"No, no, nothing that grandiose. There's going to be a special session for junior high school students, and she wanted me to give a talk. She was originally scheduled to do it, but said something came up."

"Wow. So what are you going to talk about?"

"Well, Miruka was going to present something about discrete mathematics, but I don't think I could possibly be an adequate substitute. So I told her I couldn't do it, but..."

"Let me guess. She wants you to do it anyway?"

"Of course. She says there's a university professor who's going to handle the hard stuff, but that she wants me to give a talk about something that's easier to get a grasp on, that it would be good for them to see a girl near their age talking about math. She says I can talk about anything, so long as it's related to math and information science."

"Sounds like a great opportunity. You should do it."

"Great, now I'm even getting pressured by you. But I don't see how I could—I get nervous just imagining it. I mean, there's going to be like twenty people there!"

"Twenty people isn't all that many. You should talk about all that algorithm stuff you've been studying recently. Code walkthroughs, asymptotic analysis, searching and sorting, that kind of thing."

"Now you sound even more like Miruka. That's exactly what she suggested!"

"Hey, I know. How about you ask Mr. Muraki for advice?"

Tetra sighed. "Been there, done that. He said the same thing."

"Might be time to give in, then. After all, you said you want to be the one sending messages, right?"

"Huh?"

"Seems to me that giving a presentation is a great way of sending messages."

"Yeah. Yeah, I guess so. Never really thought about it like that."

The warning bell rang.

"Oops, better get to class!" Tetra said. "Thanks for the advice. I...I'll think about it."

She gave a quick bow and bounced off to her classroom.

8.4 In the Library

8.4.1 The Wandering Problem

After my afternoon classes ended, I headed to the library as usual. Tetra was already there, studying.

"Hey!" she said, waving at me.

"You again?" I joked.

"Me again," she said, framing her face with her hands. "Oh, by the way, I found out the name of $a + d$! It's called the trace! I'm not sure *why* it's called that, but that's its name."

$$A^2 - \underbrace{(a+d)}_{\text{trace}}A + (\underbrace{ad - bc}_{\text{determinant}})E = O$$

"Interesting. Did you look that up?"

"I did! I decided to be proactive, instead of waiting for you guys to tell me."

"Well good for you. And is that another card from Mr. Muraki?"

I pointed at a card sitting on the table near her.

"It is indeed. I told him I was getting pretty comfortable with matrices, and he laid this on me."

> **Problem 8-2 (Wandering Alice)**
>
> Each year, Alice wanders between two countries A and B. In year 0 she flips a fair coin, and lives in country A if the result is heads or in country B if the result is tails. Each year after that, she flips a coin from the country she is living in, and if the result is heads she stays in that country, or if the result is tails she moves to the other country and lives there for one year.
>
> - There is a $\frac{1}{2}$ probability of heads and a $\frac{1}{2}$ probability of tails in the year 0 coin flip.
>
> - There is a $1-p$ probability of heads and a p probability of tails when flipping the country A coin.
>
> - There is a $1-q$ probability of heads and a q probability of tails when flipping the country B coin.
>
> - $0 < p < 1$ and $0 < q < 1$.
>
> Find the probability that Alice will be living in country A in year n.

"Huh," I said. "How far did you get?"

"Not far at all. I've just been wondering if we couldn't use a general term in a geometric sequence. Start with c as the first term, then make a geometric sequence with common ratio of r, like $c, cr, cr^2, cr^3, \ldots, cr^n, \ldots$ Doesn't that somehow feel right?"

"Hmm... I'm not sure..."

"No? I was thinking that then we could organize the trips back and forth between the two countries like this."

$$
\begin{array}{ccc}
A & \xrightarrow{1-p} & A \\
A & \xrightarrow{p} & B \\
B & \xrightarrow{1-q} & B \\
B & \xrightarrow{q} & A
\end{array}
$$

"I think it might be better to look at it like this," I said.

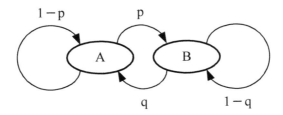

"Oh, sure," Tetra said. "That would make the movement easier to see. So what do we do next? I guess we could consider n to be the number of coin tosses, and m to be the number of heads... but wow, the number of possible combinations would get out of hand fast."

"Did you try creating an example?"

"Uh... no."

"Then you might be getting ahead of yourself."

"Good point. I guess I was just so sure that a geometric progression would be involved, so I jumped to general terms. You're right, I should have started with something more concrete."

"Right. Examples are key to understanding, remember. Instead of going straight to the nth year, start with year 0, then year 1 and so on. For example, the probability of living in country A in year 0 is $\frac{1}{2}$, right? So what's the probability that she lives in A in year 0, and stays there in year 1?"

"One step at a time, right? Let's see... There was a $\frac{1}{2}$ probability that she ended up in A in the first place, then a $1 - p$ probability that she stays there in year 1, so I guess we have to multiply those two probabilities together."

"Okay, and what's the probability that she lives in B in year 0 and then A in year 1?"

"Same kind of thing, right? The probability of moving from B to A is q, so we would multiply $\frac{1}{2} \times q$."

"So we have the probability of living in A in both years 0 and 1, and the probability of living in B in year 0, then A in year 1, so is that all we need?"

"All of what?"

"Have we covered all the situations in which Alice might live in A in year 1, with no leaks and no dupes?"

"Oh, right. Yes. Yes, I think so."

"Something wrong?"

"No, I was just thankful that you asked me to make sure about the leaks and dupes. I need to learn how to do that myself."

"It is a good habit to get into. So anyway, to make the generalization easier let's let a_n and b_n be the probabilities of living in country A or B in year n. Then we can write this for $n = 0$."

$$\text{Probability of living in country A in year 0} = a_0 = \frac{1}{2}$$

$$\text{Probability of living in country B in year 0} = b_0 = \frac{1}{2}$$

"So now we get this."

$$\text{Probability of living in country A in year 0, A in year 1} = \frac{1}{2} \times (1-p) = (1-p)a_0$$

$$\text{Probability of living in country B in year 0, A in year 1} = \frac{1}{2} \times q = qb_0$$

"All I'm doing here is replacing the one-halves with a_0 or b_0. Then we can use a_0 and b_0 to write a_1, and the probability of living in A in year 1 is the sum of $(1-p)a_0$ and qb_0."

$$a_1 = (1-p)a_0 + qb_0 \qquad \text{Probability of living in A in year 1}$$

"I see!" Tetra said. "And we can do the same thing for b_1, like this!"

$$b_1 = (1-q)b_0 + pa_0 \qquad \text{Probability of living in B in year 1}$$

"That's right," I said, "but you should write the a_0 first."

$$b_1 = pa_0 + (1-q)b_0$$

"Why's that?"

"You'll see when we line up a_1 and b_1."

$$\begin{cases} a_1 = (1-p)a_0 + qb_0 \\ b_1 = pa_0 + (1-q)b_0 \end{cases}$$

"Mmm... not sure I'm seeing it. What am I looking for?"

"One word—*matrices*."

$$\begin{pmatrix} a_1 \\ b_1 \end{pmatrix} = \begin{pmatrix} 1-p & q \\ p & 1-q \end{pmatrix} \begin{pmatrix} a_0 \\ b_0 \end{pmatrix}$$

"Hey, look at that! It's our dear old friend, multiply-multiply-add! A sum of products!"

$$\underbrace{\overbrace{(1-p)\cdot a_0}^{\text{multiply}} + \overbrace{q \cdot b_0}^{\text{multiply}}}_{\text{add}} \qquad \underbrace{\overbrace{p\cdot a_0}^{\text{multiply}} + \overbrace{(1-q)\cdot b_0}^{\text{multiply}}}_{\text{add}}$$

"That's great, but let's use our inside voice here," I said.

"Oops! Sorry!"

"Onward, then. Now that we know the relation between years 0 and 1, we can do something similar to think of the relation between years n and $n+1$."

"Okay, sure. Let me see if I can do better this time."

$$\begin{pmatrix} a_{n+1} \\ b_{n+1} \end{pmatrix} = \begin{pmatrix} 1-p & q \\ p & 1-q \end{pmatrix} \begin{pmatrix} a_n \\ b_n \end{pmatrix}$$

8.4.2 The Meaning of A^2

It was still raining outside. We continued our math talk in the quiet library.

"The vector $\begin{pmatrix} a_n \\ b_n \end{pmatrix}$ shows the probabilities of living in country A or B in year n," I said. "We can call this a probability vector for the nth year."

"Probability vector..." Tetra said, writing in her notebook.

"The product of the matrix $\begin{pmatrix} 1-p & q \\ p & 1-q \end{pmatrix}$ in the Wandering Alice problem and the probability vector $\begin{pmatrix} a_n \\ b_n \end{pmatrix}$ for the nth year gives us the probability vector for year $(n+1)$."

$$\begin{pmatrix} 1-p & q \\ p & 1-q \end{pmatrix} \begin{pmatrix} a_n \\ b_n \end{pmatrix} = \begin{pmatrix} a_{n+1} \\ b_{n+1} \end{pmatrix}$$

"Right."

"So what would this represent?"

$$\begin{pmatrix} 1-p & q \\ p & 1-q \end{pmatrix}^2 \begin{pmatrix} a_n \\ b_n \end{pmatrix} = ?$$

"You're squaring the matrix now? Hmm, I'm not sure what that would be."

"Let's calculate it out and see what happens."

$$\begin{pmatrix} 1-p & q \\ p & 1-q \end{pmatrix}^2 \begin{pmatrix} a_n \\ b_n \end{pmatrix} = \begin{pmatrix} 1-p & q \\ p & 1-q \end{pmatrix} \begin{pmatrix} 1-p & q \\ p & 1-q \end{pmatrix} \begin{pmatrix} a_n \\ b_n \end{pmatrix}$$

$$= \begin{pmatrix} 1-p & q \\ p & 1-q \end{pmatrix} \begin{pmatrix} a_{n+1} \\ b_{n+1} \end{pmatrix}$$

$$= \begin{pmatrix} a_{n+2} \\ b_{n+2} \end{pmatrix}$$

"Oh, I get it! The squared matrix lets us find the probability vector for the second year!"

8.4.3 The nth Power of a Matrix

"Okay, so let's take one more look at Mr. Muraki's card," I said. "What we're after is the probability vector for year n. In other words, $\begin{pmatrix} a_n \\ b_n \end{pmatrix}$. So what we want to do is find the nth power of the matrix. The reason why we want to do that is because if we multiply the matrix $\begin{pmatrix} 1-p & q \\ p & 1-q \end{pmatrix}^n$ by $\begin{pmatrix} a_0 \\ b_0 \end{pmatrix}$, we can get the probability vector we're after."

"Seems pretty straightforward," Tetra said. She fiddled with the card as she mulled this.

I found myself conflicted. I knew an interesting way of finding the nth power of a matrix, but then again hammering through the matrix calculations and using mathematical induction might be the more comfortable approach for Tetra.

Which way to go...

I closed my eyes as I weighed the pros and cons of each approach, and noticed that I had started tracing a circle with a finger, just like Miruka often did. I wondered if she too did that while trying to decide how to proceed with an explanation.

"Oh, wait!" Tetra said. "There's more!"

She showed me two equations written on the back of Mr. Muraki's card.

Hints:

$$\begin{pmatrix} \alpha & 0 \\ 0 & \beta \end{pmatrix}^n = \begin{pmatrix} \alpha^n & 0 \\ 0 & \beta^n \end{pmatrix} \qquad (PDP^{-1})^n = PD^n P^{-1}$$

As usual, Mr. Muraki was way ahead of me; his hint for Tetra decided which way to go with this problem.

Time to learn about matrix diagonalization.

8.4.4 Some Prep: Diagonal Matrices

"Okay," I said. "Let me show you how to use diagonalization to find the nth power of a matrix."

"You bet," Tetra said, nodding.

"As preparation, let's talk about the properties of a diagonal matrix, in other words one that looks like this."

$$\begin{pmatrix} \alpha & 0 \\ 0 & \beta \end{pmatrix}$$

"The α and β here are reals, and they make a diagonal line from upper left to lower right. All other elements are zeros. One reason why diagonal matrices like this are really nice because it's super easy to find their powers. For example, let's square this matrix."

$$\begin{pmatrix} \alpha & 0 \\ 0 & \beta \end{pmatrix}^2 = \begin{pmatrix} \alpha & 0 \\ 0 & \beta \end{pmatrix} \begin{pmatrix} \alpha & 0 \\ 0 & \beta \end{pmatrix}$$
$$= \begin{pmatrix} \alpha \cdot \alpha + 0 \cdot 0 & \alpha \cdot 0 + 0 \cdot \beta \\ 0 \cdot \alpha + \beta \cdot 0 & 0 \cdot 0 + \beta \cdot \beta \end{pmatrix}$$
$$= \begin{pmatrix} \alpha^2 & 0 \\ 0 & \beta^2 \end{pmatrix}$$

"In other words, we get a formula like this."

$$\begin{pmatrix} \alpha & 0 \\ 0 & \beta \end{pmatrix}^2 = \begin{pmatrix} \alpha^2 & 0 \\ 0 & \beta^2 \end{pmatrix}$$

"And of course we can generalize this. You can do a proof by mathematical induction later, if you want."

$$\begin{pmatrix} \alpha & 0 \\ 0 & \beta \end{pmatrix}^n = \begin{pmatrix} \alpha^n & 0 \\ 0 & \beta^n \end{pmatrix}$$

"So the nth power of a diagonal matrix is the matrix with its elements raised to the nth power?" Tetra said.

"Exactly. And that's the first bit of preparation that we need to talk about diagonalization."

"What else do we need?"

"Sandwiches."

8.4.5 More Prep: Matrix and Inverse Matrix Sandwiches

"Specifically," I said, "we need to talk about sandwiches made from a matrix and its inverse. Say we have a matrix P and its inverse P^{-1}. We're going to make our sandwich by putting these on either side of some other matrix D, and multiply the three together, like this."

$$PDP^{-1}$$

"Okay," Tetra said, "but why would we do that?"

"Because interesting things happen. Like, if we raise this PDP^{-1} to the nth power, the result is D^n sandwiched by P and P^{-1}. Let's take a look at a squared sandwich as an example, to see how the squared sandwich becomes a sandwich of a square."

$$\begin{aligned}
\text{Squared sandwich} &= (PDP^{-1})^2 \\
&= (PDP^{-1})(PDP^{-1}) \\
&= PDP^{-1}PDP^{-1} \\
&= PD(P^{-1}P)DP^{-1} \\
&= PDEDP^{-1} \qquad E = \text{identity matrix } \begin{pmatrix} 1 & 0 \\ 0 & 1 \end{pmatrix} \\
&= PDDP^{-1} \\
&= PD^2P^{-1} \\
&= \text{sandwich of a square}
\end{aligned}$$

"And the same thing happens when raising to an nth power."

$$(PDP^{-1})^n = PD^nP^{-1}$$

"We could prove this using mathematical induction too?" Tetra asked.

"Sure. The key here is that the $P^{-1}P$ becomes the identity matrix E and disappears. That's where we see that the square of a sandwich is a sandwich of a square."

8.4.6 To Eigenvalues

"Okay, we're ready to get started," I said. "We have two hints from Mr. Muraki."

▷ The nth power of a diagonal matrix is the matrix with its elements raised to the nth power

$$\begin{pmatrix} \alpha & 0 \\ 0 & \beta \end{pmatrix}^n = \begin{pmatrix} \alpha^n & 0 \\ 0 & \beta^n \end{pmatrix}$$

▷ The nth power of a matrix–inverse sandwich is a sandwich of the nth power

$$(PDP^{-1})^n = PD^nP^{-1}$$

"With all this, we can finally find the nth power of the matrix that came up for the Wandering Alice problem. Let's call that matrix A."

$$A = \begin{pmatrix} 1-p & q \\ p & 1-q \end{pmatrix}$$

"Here's what we want to do."

> Find a diagonal matrix D and matrix P fulfilling
> $A = PDP^{-1}$

"Remind me why that's the goal?" Tetra said.
"Well, let's set this up more clearly, in the form of a problem."

Problem 8-3 (Diagonalization of matrices)

Given a matrix $A = \begin{pmatrix} 1-p & q \\ p & 1-q \end{pmatrix}$, find diagonal matrix $D = \begin{pmatrix} \alpha & 0 \\ 0 & \beta \end{pmatrix}$ and matrix $P = \begin{pmatrix} a & b \\ c & d \end{pmatrix}$ fulfilling $A = PDP^{-1}$.

"So our goal comes from this relation."

$$A^n = (PDP^{-1})^n = PD^nP^{-1}$$

"In other words, we're getting the nth power of matrix A from the nth power of the diagonal matrix D. We start with $A = PDP^{-1}$."

$$A = PDP^{-1}$$
$$AP = PD \quad \text{multiply P from the right}$$

"Let's look at the components of $AP = PD$."

$$\begin{pmatrix} 1-p & q \\ p & 1-q \end{pmatrix} \begin{pmatrix} a & b \\ c & d \end{pmatrix} = \begin{pmatrix} a & b \\ c & d \end{pmatrix} \begin{pmatrix} \alpha & 0 \\ 0 & \beta \end{pmatrix}$$
$$= \begin{pmatrix} \alpha a & \beta b \\ \alpha c & \beta d \end{pmatrix}$$
$$= \alpha \begin{pmatrix} a & 0 \\ c & 0 \end{pmatrix} + \beta \begin{pmatrix} 0 & b \\ 0 & d \end{pmatrix}$$

"If we look at the $\binom{a}{c}$ in the first column of $\left(\begin{smallmatrix} a & 0 \\ c & 0 \end{smallmatrix}\right)$, we can see that this is true."

$$\begin{pmatrix} 1-p & q \\ p & 1-q \end{pmatrix} \begin{pmatrix} a \\ c \end{pmatrix} = \alpha \begin{pmatrix} a \\ c \end{pmatrix}$$

"We can also write it like this."

$$\begin{pmatrix} 1-p & q \\ p & 1-q \end{pmatrix} \begin{pmatrix} a \\ c \end{pmatrix} = \begin{pmatrix} \alpha & 0 \\ 0 & \alpha \end{pmatrix} \begin{pmatrix} a \\ c \end{pmatrix}$$

"Let's move everything to the left..."

$$\begin{pmatrix} 1-p & q \\ p & 1-q \end{pmatrix} \begin{pmatrix} a \\ c \end{pmatrix} - \begin{pmatrix} \alpha & 0 \\ 0 & \alpha \end{pmatrix} \begin{pmatrix} a \\ c \end{pmatrix} = \begin{pmatrix} 0 \\ 0 \end{pmatrix}$$

"...to get this equation."

$$\begin{pmatrix} 1-p-\alpha & q \\ p & 1-q-\alpha \end{pmatrix} \begin{pmatrix} a \\ c \end{pmatrix} = \begin{pmatrix} 0 \\ 0 \end{pmatrix}$$

"Now the question is, does the matrix $\left(\begin{smallmatrix} 1-p-\alpha & q \\ p & 1-q-\alpha \end{smallmatrix}\right)$ here have an inverse?"

"Hang on," Tetra said. "Where did that come from?"

"When you're dealing with matrices, you always want to know if they have an inverse."

"Oh, okay. So... does it?"

"No. No it doesn't."

"Oops. And how do you know that?"

"Well, since the matrix P has an inverse, we know that $\binom{a}{c} \neq \binom{0}{0}$, right?"

"Uh, we do?"

"Sure. Think about what would happen if they *were* equal. Then the determinant of P would be 0, like this."

$$|P| = \begin{vmatrix} a & b \\ c & d \end{vmatrix} = \begin{vmatrix} 0 & b \\ 0 & d \end{vmatrix} = 0 \cdot d - b \cdot 0 = 0$$

"Poof, the determinant of P has disappeared. That's a contradiction, so $\binom{a}{c} \neq \binom{0}{0}$. In other words, at least one of a or c must be nonzero."

"Okay, I'm getting this now. It all comes from the definitions."

"Back to the problem. Do you see what we can learn from this?"

$$\begin{pmatrix} 1-p-\alpha & q \\ p & 1-q-\alpha \end{pmatrix} \begin{pmatrix} a \\ c \end{pmatrix} = \begin{pmatrix} 0 \\ 0 \end{pmatrix}, \quad \begin{pmatrix} a \\ c \end{pmatrix} \neq \begin{pmatrix} 0 \\ 0 \end{pmatrix}$$

"Honestly, I'm not sure."

"Well, like I was saying, we want to ask whether there's an inverse matrix. And we know that the matrix $\begin{pmatrix} 1-p-\alpha & q \\ p & 1-q-\alpha \end{pmatrix}$ doesn't have one."

"Sorry to keep saying this, but how do we know that?"

"It's just like before. If $\begin{pmatrix} 1-p-\alpha & q \\ p & 1-q-\alpha \end{pmatrix}$ did have an inverse, then we could start with this..."

$$\begin{pmatrix} 1-p-\alpha & q \\ p & 1-q-\alpha \end{pmatrix} \begin{pmatrix} a \\ c \end{pmatrix} = \begin{pmatrix} 0 \\ 0 \end{pmatrix}$$

"...then multiply both sides by the inverse $\begin{pmatrix} 1-p-\alpha & q \\ p & 1-q-\alpha \end{pmatrix}^{-1}$, which would give us $\begin{pmatrix} a \\ c \end{pmatrix} = \begin{pmatrix} 0 \\ 0 \end{pmatrix}$."

"Oh, the same contradiction as before."

"Right, it contradicts with the $\begin{pmatrix} a \\ c \end{pmatrix} \neq \begin{pmatrix} 0 \\ 0 \end{pmatrix}$ that we started out with."

"Sure, I see that," Tetra said, grimacing. "I just wish I could see how to put all this together so quick like you do, whipping out the contradictions and all."

"Sorry for burning through the explanation so fast, but don't let that get you down. I'm only able to because I've done it myself so many times. The speed comes only through practice."

"Of course. I'll be sure to practice this too!"

"Just be sure you're practicing, not memorizing. Don't try to remember *how* things change, remember the logic by which they change."

"Like learning a story, not the words!"

"Absolutely. Okay, back to the calculations. We know that the determinant of matrix $\begin{pmatrix} 1-p-\alpha & q \\ p & 1-q-\alpha \end{pmatrix}$ is 0."

$$\begin{vmatrix} 1-p-\alpha & q \\ p & 1-q-\alpha \end{vmatrix} = 0$$

"Now we can use the definition of the determinant..."

$$\begin{vmatrix} 1-p-\alpha & q \\ p & 1-q-\alpha \end{vmatrix} = (1-p-\alpha)(1-q-\alpha) - pq$$
$$= 1 - q - \alpha - p + pq + p\alpha - \alpha + q\alpha + \alpha^2 - pq$$

"... and clean up the α's."

$$= \alpha^2 - (1-p+1-q)\alpha + (1-p-q)$$

"I'm focusing on the α's here because I'm interested in a story in which we find that value. Since the determinant equals 0, we can find α like this."

$$\alpha^2 - (1-p+1-q)\alpha + (1-p-q) = 0$$

"If we let $x = \alpha$, then we have a quadratic equation like this."

$$x^2 - (1-p+1-q)x + (1-p-q) = 0$$

"We call this the characteristic equation for the matrix $\begin{pmatrix} 1-p & q \\ p & 1-q \end{pmatrix}$. Interestingly, the characteristic equation for $A = \begin{pmatrix} 1-p & q \\ p & 1-q \end{pmatrix}$ is in the same form as the Cayley–Hamilton theorem."

$$A^2 - (1-p+1-q)A + (1-p-q)E = O \quad \text{Cayley–Hamilton theorem}$$
$$x^2 - \underbrace{(1-p+1-q)}_{\text{trace}} x + \underbrace{(1-p-q)}_{\text{determinant}} = 0 \quad \text{characteristic equation}$$

"When the characteristic equation we created from A has two solutions, they become the α, β that create the diagonal matrix D. Anyway, here's the characteristic equation."

$$x^2 - (1-p+1-q)x + (1-p-q) = 0$$

"We can factor it like this..."

$$(x-1)(x-(1-p-q)) = 0$$

"...which we solve to get this."

$$x = 1,\ 1-p-q$$

"These two values are called the 'eigenvalues' of the matrix $\begin{pmatrix} 1-p & q \\ p & 1-q \end{pmatrix}$. Now if we let $\alpha = 1$ and $\beta = 1 - p - q$, we get this diagonal matrix."

$$D = \begin{pmatrix} \alpha & 0 \\ 0 & \beta \end{pmatrix} = \begin{pmatrix} 1 & 0 \\ 0 & 1-p-q \end{pmatrix}$$

"We could also reverse α and β and get $D = \begin{pmatrix} 1-p-q & 0 \\ 0 & 1 \end{pmatrix}$, but we need to settle on one to proceed."

8.4.7 Eigenvectors

"Okay," I said. "We have the diagonal matrix D. Now we just need to find the matrix P. To do that, we need to find something called the 'eigenvector.'"

"The eigenvector..." Tetra said, adding this to a list in her notes after "characteristic equation" and "eigenvalue."

"I'm not sure how well I'll remember this," she said, "but at least I'll have a list of terms to look up to remind myself."

"The flow of the logic isn't all that hard," I said. "Just don't get lost as you follow the path."

"You bet," she said.

"So we're given a p and a q," I continued, "and we've already found α and β. Now we just need to find a, b, c, d."

$$\begin{cases} \begin{pmatrix} 1-p-\alpha & q \\ p & 1-q-\alpha \end{pmatrix} \begin{pmatrix} a \\ c \end{pmatrix} = \begin{pmatrix} 0 \\ 0 \end{pmatrix} \\ \begin{pmatrix} 1-p-\beta & q \\ p & 1-q-\beta \end{pmatrix} \begin{pmatrix} b \\ d \end{pmatrix} = \begin{pmatrix} 0 \\ 0 \end{pmatrix} \end{cases}$$

"Here, p and q are known values, and a through d are unknowns. Let's let $\alpha = 1$ and $\beta = 1 - p - q$, giving us this."

$$\begin{cases} \begin{pmatrix} -p & q \\ p & -q \end{pmatrix} \begin{pmatrix} a \\ c \end{pmatrix} = \begin{pmatrix} 0 \\ 0 \end{pmatrix} \\ \begin{pmatrix} q & q \\ p & p \end{pmatrix} \begin{pmatrix} b \\ d \end{pmatrix} = \begin{pmatrix} 0 \\ 0 \end{pmatrix} \end{cases}$$

"After some calculations and cleaning up, we'll get this system of equations."

$$\begin{cases} pa - qc = 0 \\ b + d = 0 \end{cases}$$

"That looks like a tough one to solve," Tetra said. "So many variables!"

"Yeah, four variables a, b, c, d, but only two equations, which means we can't completely nail down values for the variables. But that's okay—we don't need to. We just need to put together matrices P and D such that $A = PDP^{-1}$. For example, this fits in the system of equations."

$$a = q, \ b = -1, \ c = p, \ d = 1$$

"That lets us construct the matrix P."

$$P = \begin{pmatrix} a & b \\ c & d \end{pmatrix} = \begin{pmatrix} q & -1 \\ p & 1 \end{pmatrix}$$

"The inverse for $P = \begin{pmatrix} a & b \\ c & d \end{pmatrix}$ is $P^{-1} = \frac{1}{ad-bc}\begin{pmatrix} d & -b \\ -c & a \end{pmatrix}$, so when $a = q, b = -1, c = p, d = 1$, the inverse P^{-1} is this."

$$\begin{aligned}
P^{-1} &= \frac{1}{ad - bc} \begin{pmatrix} d & -b \\ -c & a \end{pmatrix} \\
&= \frac{1}{q \cdot 1 - (-1) \cdot p} \begin{pmatrix} 1 & -(-1) \\ -p & q \end{pmatrix} \\
&= \frac{1}{p + q} \begin{pmatrix} 1 & 1 \\ -p & q \end{pmatrix}
\end{aligned}$$

> **Answer 8-3 (Diagonalization of matrices)**
>
> For matrix $A = \begin{pmatrix} 1-p & q \\ p & 1-q \end{pmatrix}$,
>
> $$D = \begin{pmatrix} 1 & 0 \\ 0 & 1-p-q \end{pmatrix}$$
>
> $$P = \begin{pmatrix} q & -1 \\ p & 1 \end{pmatrix}$$
>
> $$P^{-1} = \frac{1}{p+q}\begin{pmatrix} 1 & 1 \\ -p & q \end{pmatrix}$$
>
> satisfies $A = PDP^{-1}$.

8.4.8 Finding A^n

"Okay, we're finally ready to find A^n," I said. "To do so, all we need to do is combine everything we've done so far. Namely, this."

$$\begin{cases} D &= \begin{pmatrix} 1 & 0 \\ 0 & 1-p-q \end{pmatrix} \\ P &= \begin{pmatrix} q & -1 \\ p & 1 \end{pmatrix} \\ P^{-1} &= \dfrac{1}{p+q}\begin{pmatrix} 1 & 1 \\ -p & q \end{pmatrix} \end{cases}$$

"From that, we get this."

$$A^n = (PDP^{-1})^n$$
$$= PD^n P^{-1}$$
$$= \begin{pmatrix} q & -1 \\ p & 1 \end{pmatrix} \begin{pmatrix} 1 & 0 \\ 0 & 1-p-q \end{pmatrix}^n \cdot \frac{1}{p+q} \begin{pmatrix} 1 & 1 \\ -p & q \end{pmatrix}$$
$$= \begin{pmatrix} q & -1 \\ p & 1 \end{pmatrix} \begin{pmatrix} 1^n & 0 \\ 0 & (1-p-q)^n \end{pmatrix} \cdot \frac{1}{p+q} \begin{pmatrix} 1 & 1 \\ -p & q \end{pmatrix}$$
$$= \begin{pmatrix} q & -(1-p-q)^n \\ p & (1-p-q)^n \end{pmatrix} \cdot \frac{1}{p+q} \begin{pmatrix} 1 & 1 \\ -p & q \end{pmatrix}$$
$$= \frac{1}{p+q} \begin{pmatrix} q+p(1-p-q)^n & q-q(1-p-q)^n \\ p-p(1-p-q)^n & p+q(1-p-q)^n \end{pmatrix}$$

"We can use this to find the probability vector for year n."

$$A^n \begin{pmatrix} a_0 \\ b_0 \end{pmatrix} = A^n \begin{pmatrix} \frac{1}{2} \\ \frac{1}{2} \end{pmatrix}$$
$$= \frac{1}{p+q} \begin{pmatrix} q+p(1-p-q)^n & q-q(1-p-q)^n \\ p-p(1-p-q)^n & p+q(1-p-q)^n \end{pmatrix} \begin{pmatrix} \frac{1}{2} \\ \frac{1}{2} \end{pmatrix}$$
$$= \frac{1}{2(p+q)} \begin{pmatrix} q+p(1-p-q)^n + q-q(1-p-q)^n \\ p-p(1-p-q)^n + p+q(1-p-q)^n \end{pmatrix}$$
$$= \frac{1}{2(p+q)} \begin{pmatrix} 2q+(p-q)(1-p-q)^n \\ 2p-(p-q)(1-p-q)^n \end{pmatrix}$$

"So now we can find this."

$$\begin{cases} a_n = \frac{1}{2(p+q)}(2q+(p-q)(1-p-q)^n) = \frac{q}{p+q} + \frac{p-q}{2(p+q)}(1-p-q)^n \\ b_n = \frac{1}{2(p+q)}(2p-(p-q)(1-p-q)^n) = \frac{p}{p+q} - \frac{p-q}{2(p+q)}(1-p-q)^n \end{cases}$$

"And the probability that we're after is the a_n one, right?"

Answer 8-2 (Wandering Alice)

The probability that Alice lives in country A in year n is

$$\frac{q}{p+q} + \frac{p-q}{2(p+q)}(1-p-q)^n.$$

"Wow," Tetra said. "I think I understand all the parts, but the whole..."

"Here, let me draw out a map of our journey to matrix diagonalization. There's lots of letters, but it might help keep track of what we're using to find what."

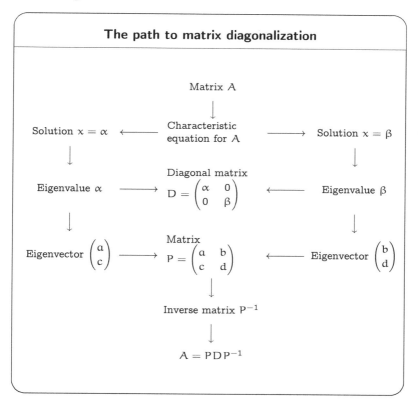

"Wow, that was quite a trip. So can we use this matrix diagonalization to find the nth power of any matrix?"

"If I remember correctly things are a little different when the characteristic equations have repeated roots."

"Oh, okay. In any case, that seems like a really complex answer to a pretty simple question."

"Where will someone be living in what year? Yeah, I agree. If you think about it, Alice's actions really only depend on where she was in the previous year. The problem doesn't have any kind of 'memory' about where she was at any time before that. It feels like this problem could be a simplification of some problem from the social sciences. Maybe more, some scientific experiment... Hey, wait a minute! This is a random walk!"

"A what?"

"A random walk. The two countries, they represent two states that can be wandered among. This is a kind of random walk problem!"

8.5 AT HOME

8.5.1 Apprehension (3)

That evening I was studying alone in my room. Well, I was trying to at least. I had spent the day talking about math with Tetra, but to tell the truth it was Miruka who was on my mind.

I had solved the Wandering Alice problem using diagonalization of matrices. If I'd had more time, I likely would have been able to continue pushing on to even more interesting places. For example, what if instead of only two countries—or conditions, or however you want to look at it—there were three? What would happen as the number of possibilities increased to infinity? And so on. I had no doubt that the more I thought about it, the more I would reveal new problems and interesting discoveries.

Even so, I wished Miruka had been there. With her knowledge and creativity, who knows what new worlds she would have led us to? Even if she used the same method that I had, I was sure she could have taken it farther, higher.

Miruka's absence made her existence all the more noticeable.

With my head full of such thoughts, I headed toward the kitchen to see what was for dinner. The phone rang as I passed by, as if it had been waiting for me.

"Hello?"

"Do promises scare you?" said the voice on the other end.

"Is that Miruka? Er, good evening, I guess?"

"It's early morning here."

"Where's here?"

"In the States, on the West Coast. I'll be here for a week."

"The States? As in, the United States? What the—?"

"You avoid making promises."

"What are you talking about?"

"About how promises scare you."

"I— Wait, seriously? You're calling me from America?"

"Why do they scare you?"

"I, uh, I'm afraid I'll break them, I guess? This is an expensive call, right?"

"A promise is an expression of intent."

"Intent?"

"The path you intend to walk. So yours will avoid all promises, out of fear you might break them? You want to be remembered as someone who never broke a promise, because you never made one? That's the life you're aiming at?"

I remained silent, unsure how to respond.

"Is that it? A life of peace, only because nothing is gained or lost?" Miruka pressed on. "So what is it? What's your path?"

"I ... I don't ... "

"Breaking a promise is bad. Being unable to live up to a promise is unfortunate. But never making a promise? That's just cowardice."

And with that she hung up.

8.5.2 A Rainy Night

After Miruka's phone call, I felt as if my brains had been scrambled. Maybe she had called me on some whim, but there was nothing whimsical about the question she had left me with—what path did I intend to walk? It was a good question, and one that I needed to face. But was it one worth a sudden international telephone call?

Hang on a minute...

I did some mental calculations and realized that it was around 2:00 AM on the US West Coast. It wasn't early morning there, it was the middle of the night.

What could be keeping her up?

Yuri was worried about her letter, Tetra about her first conference presentation, and I was as confused as ever. And now even Miruka seemed uneasy. All of us, so anxious about our future.

There was still something strange about thinking of Miruka as being in a defensive posture, despite what I had seen on the riverbank. She still wasn't a person I thought of as needing someone to call in the middle of the night. That was silly, of course—being good at solving problems had nothing to do with protecting one's own feelings—but still...

I opened a window. A cold, moist scent of rain filled the kitchen. It was getting dark outside. I peered into the gloom, as if searching for my own future.

> The way a problem is solved is generally much more important than the solution itself.
>
> DONALD KNUTH
> *Selected Papers on Analysis of Algorithms*

CHAPTER **9**

Strongly, Correctly, Beautifully

> What was it to me, if when I had chosen a vast tree in the woods, and with much trouble cut it down, if I had been able with my tools to hew and dub the outside into the proper shape of a boat, and burn or cut out the inside to make it hollow, so as to make a boat of it—if, after all this, I must leave it just there where I found it, and not be able to launch it into the water?
>
> DANIEL DEFOE
> *Robinson Crusoe*

9.1 AT HOME

9.1.1 A Rainy Saturday

That Saturday afternoon I was studying in my room, trying to ignore the rain. I was so very ready for the rainy season to come to an end.

"Don't you just love the rain?!" Yuri enthused as she barged into my room.

"Didn't you say you hate it just the other day?" I said.

"All in the distant past. Today I've got a problem for you."

Yuri opened her notebook on top of mine, grinning ear to ear.

> **Problem 9-1 (The strong, correct, beautiful problem)**
>
> Can all of the following conditions be fulfilled?
>
> P1. Strong or correct or beautiful.
>
> P2. Kind or correct or not beautiful.
>
> P3. Not strong or kind or beautiful.
>
> P4. Not strong or not kind or correct.
>
> P5. Not kind or not correct or beautiful.
>
> P6. Not strong or not correct or not beautiful.
>
> P7. Strong or not kind or not beautiful.
>
> P8. Strong or kind or not correct.

"What's this?" I asked.

"Just what it looks like. It's a logic problem."

"Huh," I said, taking a closer look at the problem. "So for example to fulfill the condition 'P1. Strong or correct or beautiful' you have to be strong, or you have to be correct, or you have to be beautiful, right?"

"Right. I'm beautiful, so I fulfill condition P1."

"Okay, so we just need to fulfill at least one statement per condition. And the problem wants to know if it's possible for one person to fulfill all eight conditions at the same time."

"Think you can solve it?"

"Maybe," I said, "but strong and correct and beautiful are all subjective states. Something doesn't feel right about using things like that in a logic problem."

"I guess, but, uh..." Yuri pulled a piece of paper out of an envelope she'd been holding behind her back, and took a peek at what was written there. "Oh, right. It says to assume some appropriate definition for what all those words mean."

"Okay, well in that case I want to rewrite these as logic statements, like this."

$$\text{Strong } \lor \text{ Correct } \lor \text{ Beautiful}$$

"At least that way I feel a little bit less like I'm turning adjectives into propositions."

"Oops, I fail P6," Yuri said.

P6. Not strong or not correct or not beautiful.

"I'm strong and correct *and* beautiful!" she said.

"I see. So some of the conditions have negations, like P6. We can show that using logic symbols too."

$$\neg \text{ Strong } \lor \neg \text{ Correct } \lor \neg \text{ Beautiful}$$

I thought on the problem a little more. If we assumed Strong, then condition P1 would be fulfilled, as would conditions P7 and P8. So maybe next I should look at conditions that have 'not Strong' in them, like P3.

P3. Not strong or kind or beautiful.

If we've already established Strong, then to fulfill condition P3 we would need to have either Kind or Beautiful. So if we assume Beautiful, then we would fulfill that condition, and condition P2 as well. On to condition P4.

P4. Not strong or not kind or correct.

So if we already have both Strong and Kind, to fulfill condition P4 we would definitely need Correct. Next, condition P5.

P5. Not kind or not correct or beautiful.

We already had Kind and Correct, so now we needed Beautiful.

Okay, I had made good progress. With Strong, Kind, Correct, and Beautiful I had cleared several of the conditions. P1 through P5 were fulfilled, as were P7 and P8. Only P6 was left.

P6. Not strong or not correct or not beautiful.

Nope. Strong–Kind–Correct–Beautiful didn't match this condition. I was starting to wonder if a solution could be found, but maybe a different route would work. I decided to start with 'not Strong.'

- If I assume 'not Strong,' then from condition P1 I needed Correct or Beautiful.
- If I assumed Correct, then from condition P8 I needed Kind.
- From condition P7 I needed 'not Beautiful.'

Okay, so let's test this with not Strong–Correct–Kind–not Beautiful.

- Condition P1 is okay, from Correct.
- Conditions P2 and P8 are okay, from Kind.
- Conditions P3, P4, and P6 are okay, from not Beautiful.
- Condition P7 is okay, from not Beautiful.
- The final condition P5... doesn't work!

> P5. Not kind or not correct or beautiful.

"Yuri, I'm not sure there's any way to fulfill all eight conditions," I said.

"Oh yeah? Prove it," she said. "Like you always say, without a proof, you're just guessing."

With that, Yuri hit me where it hurt.

"A proof, huh? So I guess we have to decide a yes/no for each of Strong, Correct, and Beautiful, and see if we can find a combination that fits all of the conditions P1 through P8."

"You forgot Kind," Yuri said.

"Ah, right. And Kind. I guess the most straightforward approach would be to create a table."

I pulled out a sheet of graph paper and started working.

	S	C	B	K	P1	P2	P3	P4	P5	P6	P7	P8
(1)	✗	✗	✗	✗	✗	✓	✓	✓	✓	✓	✓	✓
(2)	✗	✗	✗	✓	✗	✓	✓	✓	✓	✓	✓	✓
(3)	✗	✗	✓	✗	✓	✗	✓	✓	✓	✓	✓	✓
(4)	✗	✗	✓	✓	✓	✓	✓	✓	✓	✓	✗	✓
(5)	✗	✓	✗	✗	✓	✓	✓	✓	✓	✓	✓	✗
(6)	✗	✓	✗	✓	✓	✓	✓	✓	✗	✓	✓	✓
(7)	✗	✓	✓	✗	✓	✓	✓	✓	✓	✓	✓	✗
(8)	✗	✓	✓	✓	✓	✓	✓	✓	✓	✓	✗	✓
(9)	✓	✗	✗	✗	✓	✓	✗	✓	✓	✓	✓	✓
(10)	✓	✗	✗	✓	✓	✓	✓	✗	✓	✓	✓	✓
(11)	✓	✗	✓	✗	✓	✗	✓	✓	✓	✓	✓	✓
(12)	✓	✗	✓	✓	✓	✓	✓	✗	✓	✓	✓	✓
(13)	✓	✓	✗	✗	✓	✓	✗	✓	✓	✓	✓	✓
(14)	✓	✓	✗	✓	✓	✓	✓	✓	✗	✓	✓	✓
(15)	✓	✓	✓	✗	✓	✓	✓	✓	✓	✗	✓	✓
(16)	✓	✓	✓	✓	✓	✓	✓	✓	✓	✗	✓	✓

"What's with the Xs and checkmarks?" Yuri asked as I was making my table.

"If we're assuming Strong for one condition, I put a check for that condition in the 'S' column. If we're assuming 'not Strong' instead, it's an X. Same thing for the other conditions—correct, beautiful, and kind in the C, B, and K columns. The rows here describe every possible combination of is/isn't for each of the four states, so there's sixteen in all. Then we just go down the list and see if any of them fulfill each of the conditions P1 through P8. A check in the conditions columns means that the condition is fulfilled, and X means it isn't."

"Okay, got it."

When I was done I showed the results to Yuri.

"So it looks like no matter which of the sixteen combinations you look at, each one fails one of the conditions P1 through P8. See how each row has at least one X under the P columns? We've looked at every possible combination, and none of them fulfill all eight of the conditions, so that's it. There's your proof."

"Took you long enough," Yuri said.

> **Answer 9-1 (The strong, correct, beautiful, *and kind* problem)**
>
> It is impossible to satisfy all eight conditions P1–P8.

"This is a very well-constructed problem," I said, reviewing my table. "Look how it's set up so that every combination fails exactly one condition, and every condition has a combination that fails. That means if you drop any one of the eight conditions, there would be exactly one combination that fulfills the remaining seven."

"Yeah, I noticed that! So you'd say the person who created this problem is pretty smart?"

"Sure. It's a good problem... Hang on, are you saying that you created it?"

"No, not me." Yuri glanced at the paper she had pulled out of the envelope earlier.

"Another problem?" I asked, trying to get a look.

"No!" Yuri said, whisking the paper away. "This isn't for you."

"Sheesh, what's up with you?" I said. Then I realized what was up. "Ah. You finally got an answer."

"Maybe. Okay, yeah. Still, how can it take so long for someone to write a letter? I mean seriously. And most of it's just a logic problem! I just—" Yuri stopped, and her cheeks turned red.

"Hey, Yuri."

"Huh?"

"I'm glad you finally heard back from him."

"Yeah, me too."

9.2 IN THE LIBRARY

9.2.1 *The Logical Approach*

"Oh, that's great!" Tetra squealed.

We were in the library after class. I had just told her about Yuri's letter, and Tetra looked even happier than Yuri had.

"I figured it was probably you who recommended sending a letter," I said.

"Well that's the best way to get your feelings across. Who doesn't love seeing how someone else feels in carefully chosen words?"

"He also sent a logic problem along with his answer."

"Wow, Yuri's boyfriend sounds like a very interesting person."

"Boyfriend? I don't think—"

"Hey, look! It's Miruka!"

I looked up and saw the beautifully elegant Miruka coming through the door, red-haired Lisa at her side. As before, I was unsure whether the two were friends or enemies.

9.2.2 The Satisfiability Problem

As we talked, I told Miruka, Tetra, and Lisa about the Strong–Correct–Beautiful–Kind problem Yuri had showed me.

"A Boolean satisfiability problem," Miruka said, raising an eyebrow.

"It has a name?" I said. I had just thought of it as a simple combinations problem, so Miruka's apparent interest was surprising. "Is it a famous problem?"

"In a sense. The more generalized satisfiability problem that this one is based on leads to the most famous unresolved problem in all of computer science."

"Seriously?" Tetra said. "What's the bigger problem?"

"Say you have some logical expression. You want to know what Boolean values you can assign to its variables to make the overall expression true, if that's even possible. Maybe it isn't. Finding an efficient algorithm that answers that question is a major quest in computer science."

Everyone was quiet for several seconds.

"That doesn't seem—" Tetra said.

"I wouldn't think that—" I said.

Miruka raised a hand to stop us. "I know, I know. It doesn't sound all that hard."

I pressed on. "I mean, you're just assigning true or false values to variables. There's going to be a finite number of variables, so there's a finite number of their combinations. And determining the truth of a logic statement isn't all that hard."

"What do you think, Tetra?" Miruka asked, pointing at her.

"Pretty much the same thing. Going through all the combinations is the kind of thing computers are really good at, right?"

"Hmph. And how about you, Lisa?" Miruka asked, redirecting her pointing finger.

"Inefficient," Lisa said.

"Exactly. Yes, if you perform an exhaustive search over every possible combination you can determine whether there's an assignment that makes the statement true, and if so you can even learn what that assignment is. But the order of such an algorithm is huge. In other words, an exhaustive search is inefficient. Problem is, we still haven't found an algorithm that can do it efficiently."

"Efficiently in what sense?" I asked.

"In the sense of being able to restrain the number of execution steps to some small power of the problem size n. For example, if you let n be the number of variables, we want there to be some constant K such that we can get an answer after at most n^K steps."

"Hmm, still not sure I'm seeing it."

"Then let's get a bit more formal," Miruka said, making a "gimme" gesture.

I sighed and handed her my notebook and pencil.

9.2.3 3-SAT

"I'm going to abbreviate 'Boolean satisfiability problem' as 'SAT'," Miruka began. "By satisfiability, we specifically mean 'can a logical expression be satisfied?' But first we need to define some terms."

Miruka opened my notebook and began writing.

"So the elements making up a logic statement are variables, which can take one of two values, true or false."

$$x_1 \quad x_2 \quad x_3 \quad \text{(An example of three variables)}$$

"We can also add a 'not' symbol in front of a variable."

$$\neg \quad \text{(The 'not' operator)}$$

"The not symbol is a logical operator that reverses the truth or falsity of the variable that follows it. So if x_1 is false then $\neg x_1$ is true, and if x_1 is true then $\neg x_1$ is false."

x_1	$\neg x_1$
False	True
True	False

Truth table for the $\neg$ logical operator.

"A variable or its negation is called a 'literal'..."

$$x_1 \quad \neg x_2 \quad \neg x_3 \quad \text{(3 literals)}$$

"...and a string of literals connected by OR operators is called a 'clause.'"

$$x_1 \lor \neg x_2 \lor \neg x_3 \quad \text{(a clause)}$$

"If there's even one true literal in a clause, then the clause itself is true. The clause will be false only when every literal in it is false."

L_1	L_2	L_3	$L_1 \lor L_2 \lor L_3$
False	False	False	False
False	False	True	True
False	True	False	True
False	True	True	True
True	False	False	True
True	False	True	True
True	True	False	True
True	True	True	True

Truth table for a clause (L_1, L_2, L_3 are literals).

"You can represent some very complex situations using clauses. For example, if x_1 is Strong, x_2 is Correct, and x_3 is Beautiful, the clause $x_1 \lor \neg x_2 \lor \neg x_3$ means 'Strong OR not Correct OR not Beautiful.' The conditions P1 through P8 in your problem can be represented as eight clauses.

"When you block off some number of clauses using parentheses and tie them together using AND operators, the result is called a 'logical expression.' This expression is in what's called conjunctive normal form, CNF for short."

$$(x_1 \lor \neg x_2) \land (\neg x_1 \lor x_2 \lor x_3 \lor \neg x_4) \qquad \text{(a CNF expression)}$$

"If there's even one false clause in a CNF, the CNF itself becomes false, and if all its clauses are true the CNF is true."

C_1	C_2	$(C_1) \land (C_2)$
False	False	False
False	True	False
True	False	False
True	True	True

Truth table for a CNF (C_1, C_2 are clauses).

"A CNF in which every clause has three literals is called 3-CNF."

$$(x_1 \lor \neg x_2 \lor \neg x_3) \land (x_2 \lor x_3 \lor \neg x_4) \qquad \text{(a 3-CNF)}$$

"If we assume x_1, x_2, and x_3 keep the same meaning as before and add x_4 to represent Kind, then the 3-CNF $(x_1 \lor \neg x_2 \lor \neg x_3) \land (x_2 \lor x_3 \lor \neg x_4)$ means 'Strong OR not Correct OR not Beautiful' AND 'Correct OR Beautiful OR not Kind.' Good so far?"

"I think so," Tetra said. "But all these new terms..."

"Here's a summary."

a 3-CNF logical expression

"This logical expression has two clauses, both of which have three literals, so this expression is 3-CNF. Investigating whether there exists an assignment of variables satisfying a 3-CNF expression is called a '3-SAT' problem."

9.2.4 Satisfying

"Okay, so we've talked about variables, literals, clauses, and CNF logical expressions. A variable or its negation is a literal, literals tied together with OR operators are a clause, clauses tied together with AND operators are a CNF, and if all clauses in the CNF have three literals then it's a 3-CNF."

$$(x_1 \lor \neg x_2 \lor \neg x_3) \land (x_2 \lor x_3 \lor \neg x_4) \qquad \text{(a 3-CNF expression)}$$

"Giving each variable a true or false value is called 'assignment.' For example, here's an assignment of values to the four variables in the 3-CNF expression we've been using."

$$(x_1, x_2, x_3, x_4) = (T, T, F, F) \qquad \text{(value assignment)}$$

"This particular assignment makes that 3-CNF expression true. Generally speaking, if an assignment a makes a logical expression f true, we say that f is 'satisfied.' We can similarly talk about satisfying variables, literals, and clauses—'Satisfying the variable x_1,' 'satisfying the literal $\neg x_3$,' 'satisfying the clause $x_1 \lor \neg x_2 \lor \neg x_3$,' and so on. So what your Strong–Correct–Beautiful–Kind problem is really asking is, 'Is there an assignment that satisfies a 3-CNF expression formed from the eight clauses in conditions P1–P8?'"

9.2.5 Assignment Practice

Tetra scrambled to get a list of the new terms into her notebook. "Literal... clause... assignment..."

"Let's see if you've got everything so far," Miruka said. "Find an assignment that fulfills this 3-CNF."

$$(x_1 \lor \neg x_2 \lor \neg x_3) \land (\neg x_1 \lor x_2 \lor x_4)$$

"An assignment? So I need to assign each variable as true or false, right? Let's see... how about this?"

$$(x_1, x_2, x_3, x_4) = (T, T, T, T)$$

"Looks like so long as x_1 and x_2 are true, it doesn't really matter what x_3 and x_4 are," I said. "So there should be plenty of other assignments that work too."

"Sure," Tetra said. "Which makes it all the easier to find a solution, so—"

"That's because this one is so short," Miruka cut in. "Things won't always be so easy. So let's see if we can find an algorithm that can determine whether an assignment that satisfies an arbitrary 3-CNF is possible."

"That way we can leave it all up to a computer!"

"The simplest way to solve satisfiability problems is brute force—just trying every possibility to see what works."

"I guess that should work," Tetra said.

"The problem is efficiency. Tell me, if you're using a brute force algorithm, what's the worst-case scenario for finding an assignment that works?"

"Well, each variable can take one of two possible values," I said, "so if there are n variables there would be 2^n possible assignments."

"Which means if we have 4 variables, there would be $2^4 = 16$ possible solutions!" Tetra said.

"If there are 2^n assignments in total," Miruka said, "then the number of steps you'll have to execute would be 2^n at minimum, an exponential order."

"Right," I said. "Which means if we even had just 34 variables, there would be over ten billion assignments to look at."

"17,179,869,184 solutions," Lisa muttered.

9.2.6 NP-complete problems

"3-SAT is related to predictions of how difficult a problem will be," Miruka said. "The P versus NP conjecture."

"What do you mean, how difficult a problem will be?" I asked.

"Say you have a problem of size n. If we can solve it with a calculation time of at most some constant power of n—the problem is of order $O(n^K)$, in other words—then we say that it's solvable in polynomial time. We also say that a problem like that is of class P, or just P, for 'polynomial.' We consider P problems to be efficiently solvable, or 'fast.'"

"P problems, huh?"

"There's also something called NP problems. For those, when presented with a candidate solution we can efficiently determine whether they are indeed correct. But that doesn't necessarily mean we can quickly discover a correct solution."

"I guess NP stands for 'not polynomial'?" Tetra said.

Miruka shook her head. "A common misunderstanding, but no. The 'N' stands for 'nondeterministic.' So an NP problem is one that can be solved in nondeterministic polynomial time. To give a full explanation, we would have to talk about a hypothetical computer called a Turing machine."

"Ah."

"Let's save that for another time. For now, just trust me that it has already been proven that all P problems are also NP problems. But the reverse? Are all NP problems also P problems? No one knows. If so, we can say that $P = NP$. On the other hand, if there is even one NP problem that is not also a P problem, then $P \neq NP$."

"Interesting," I said.

"The P versus NP conjecture predicts that the set of P problems does not match the set of NP problems. Roughly put, it says that just because we can quickly validate a solution, that doesn't necessarily mean we can quickly find one. Nearly all computer scientists believe that's the case. But we don't yet have a proof."

"And if we don't have a proof, we're just guessing," I said.

"Some NP problems are also what's called NP-complete problems. In a sense, they're like even harder NP problems. If we could prove that even one NP-complete problem is also a P problem, then we would have also shown that *all* NP-complete problems are P problems, in other words that $P = NP$. NP-complete problems are thus key to taking on the P versus NP conjecture. This satisfiability problem SAT we've been talking about was the first problem shown to be NP-complete. Stephen Cook received the Turing Prize for that in 1982."

"Way to go, Stephen!" Tetra said.

"For example, if you're given a candidate assignment, you can quickly determine its satisfiability. But we haven't yet found an SAT algorithm that can quickly find a correct assignment. Maybe no such algorithm exists. Or maybe one exists, but we haven't found it yet. Nearly all computer scientists believe that no such algorithm exists, but again, nobody has given a proof either way."

"And without a proof we're just guessing," Tetra said, mirroring my words.

"We also don't have a proof for the P versus NP conjecture, so an efficient SAT algorithm isn't necessarily impossible. If someone somewhere could find such an algorithm, they would revolutionize the field of computer science. Currently, we know lots of NP problems. Like I said, if we can find an efficient algorithm for the NP-complete SAT, we will have shown that all NP problems can be quickly solved. That makes SAT a very important problem, and one that is being studied extensively."

"Wow," I said, stunned that what I had first taken to be just a simple combinations problem could have such depth.

"Actually," Miruka said, "Just the other day on the plane I was reading a paper about an algorithm for solving 3-SAT problems."

"What?" Tetra shouted. "So somebody solved the P versus NP conjecture?!"

"No, the algorithm didn't provide an efficient solution. But the paper did describe how to use a probabilistic method to lower its order."

"Probability...in an algorithm?" Tetra said.

Miruka nodded. "A kind of randomized algorithm."

"What's that?" Tetra asked, but just then—

"The library is *closed*," Ms. Mizutani announced.

9.3 On the Way Home

9.3.1 Oaths and promises

We cut through the narrow streets on the way to the station, Tetra, followed by me, Miruka, then Lisa.

"I went to a relative's wedding the other day," Tetra said, turning backward as we walked, making me worry she would stumble and fall. "The bride's white wedding dress was *so* beautiful. I literally cried. 'Each one of you also must love his wife as he loves himself, and the wife must respect her husband,' they said. And 'in sickness and in health.' Just lovely..."

"Take care of your spouse when they're sick and...when they're not?" I said.

"I think it's supposed to mean always."

Wedding vows, huh. A promise made before God and humanity. A promise...

I recalled Miruka's late-night phone call. *A promise is an expression of intent*, she had said, leaving me to wonder about my own intentions.

"Why the face?" Miruka asked.

"It's nothing," I said.

9.3.2 The Conference

We made it to the main street and waited for the signal.

"So what's your answer about the conference?" Miruka asked Tetra.

"The conference, right," Tetra mumbled. "Yeah, I don't think—"

"Conference?" Lisa said.

"The one at the Narabikura Library, of course." Miruka said. "Aren't you on the prep staff?"

"Office staff."

"It's a wonderful opportunity," I said to Tetra. "You should do it. I gave a presentation at our culture festival in junior high. I learned a lot from doing it. It's a special session for junior high students, to be delivered by a high school student, right? Lots of kids might discover the joy of algorithms, thanks to you."

Tetra paused, then shrugged. "I guess you're right. Okay, fine. I'll do it," she said with a serious expression. "If there are people willing to listen to me, I guess I owe it to them to do my best."

"Hey, I bet Yuri would love to go too."

"Oh, Yuri! This would be perfect for her!" Tetra said, clapping her hands.

"Yuri?" Lisa said.

"My cousin, third year in junior high. She likes math too. You're right, Tetra. It would be perfect for her."

"Here," Lisa said, pulling a folded paper out of her bag. It was a conference pamphlet, bearing the Narabikura Library logo and an event schedule.

"Hey, Miruka, this still has your name on it for the junior high presentation," I said.

"I wasn't able to notify them of my change in plans in time."

"Such a nuisance," Lisa muttered. I figured the sudden change must have been a hassle for the organizers.

"What will you be doing?" Tetra asked.

"I'll be back in the States again."

9.4 In the Library

9.4.1 A Randomized Algorithm for Solving 3-SAT

I went back to the library the next day after my classes.

Lisa was there, typing away on her computer. Miruka was standing behind her, giving instructions.

"And?" Lisa said.

"Print ⟨Likely unsatisfiable⟩, then return," Miruka said.

"Done."

"What are you working on?" I said, peeking at the display.

"A randomized algorithm for solving satisfiability problems," Miruka answered.

A randomized algorithm for solving 3-SAT (input and output)

Input
- Logical expression (3-CNF) f
- Number of variables n
- Number of rounds R

Output

If an assignment satisfying logic statement f can be found within R rounds,
  output ⟨Satisfiable⟩.

If no satisfying assignment for f is found,
  output ⟨Likely unsatisfiable⟩

> **A randomized algorithm for solving 3-SAT (procedure)**
>
> $W1$:   procedure RANDOM-WALK-3-SAT(f, n, R)
> $W2$:   r ← 1
> $W3$:   while r ⩽ R do
> $W4$:   a ← ⟨Randomly assign n variables⟩
> $W5$:   k ← 1
> $W6$:   while k ⩽ 3n do
> $W7$:   if ⟨assignment a satisfies f⟩ then
> $W8$:   return ⟨Satisfiable⟩
> $W9$:   end-if
> $W10$:   c ← ⟨A clause in f not satisfied by assignment a⟩
> $W11$:   x ← ⟨Randomly select a variable from clause c⟩
> $W12$:   a ← ⟨Obtain an assignment by reversing the value of variable x in assignment a⟩
> $W13$:   k ← k + 1
> $W14$:   end-while
> $W15$:   r ← r + 1
> $W16$:   end-while
> $W17$:   return ⟨Likely unsatisfiable⟩
> $W18$: end-procedure

"That looks... hard," came a voice from behind me. I turned to see Tetra standing there.

"Nah. RANDOM-WALK-3-SAT is a fun algorithm," Miruka said. "Let's talk about it."

9.4.2 Random Walks

"The RANDOM-WALK-3-SAT algorithm starts with a random assignment of n variables, and changes that assignment as it goes. We repeat a random walk like that, checking whether the assignment satisfies a logical expression f each time. Line $W4$ determines where we begin the random walk, and line $W11$ determines how we change the assignment. We make those decisions using random numbers in two loops."

- The inner loop performs a random walk of $3n$ steps.
  (Initialize variable k at line *W5*, then perform the while loop from line *W6* to line *W14*.)
- The outer loop performs the random walk over R rounds.
  (Initialize variable r at line *W2*, and perform the while loop from line *W3* to line *W16*.)

"At line *W4*, we create a random assignment of n variables, and store it into assignment a. This is the start of the random walk. At line *W7*, we check whether the current assignment a satisfies the logical expression f. At line *W11*, we determine the next step in the random walk. At line *W12*, we store a new assignment into variable a. And that's pretty much the gist of RANDOM-WALK-3-SAT. What do you think?"

Tetra frowned. "I'm not sure what to think. For starters, what's a random walk?"

"He's already drawing a picture for you," Miruka said, pointing at me.

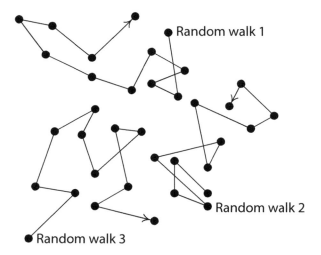

"What's this?" Tetra asked.

"The random walk that Miruka just described to us is performed over three rounds—we walk for a while, then start over from another place, then another. Right?"

"Right," Miruka said.

"I think I get the idea," Tetra said, "but what do you mean, start at a 'place'? You're trying to determine satisfiability for a logical expression, right? How are these black dots a place?"

Miruka looked at me. "Well?"

"I think each dot is an assignment," I said. "Each one is like a set of true/false values for each variable. There are n Boolean variables, so there are 2^n possible assignments. That means what we're walking over is 2^n sets of n true-or-false values."

"Okay, I get it. Except for the parts I don't get. What's the point of the random walk? Isn't the algorithm just randomly picking out assignments?"

"No," Miruka said. "The only random choice from among all the assignments comes at line $W4$. What you really need to look at is line $W10$. That's where we're picking out clauses from the logical expression that aren't satisfied."

Miruka paused, as if to give Tetra time for that to settle in. After some time looking at the procedure, Tetra nodded.

"Okay, sure. The logical expression is held together with AND operators like this..."

$$(\text{clause}_1) \wedge (\text{clause}_2) \wedge \cdots \wedge (\text{clause}_{123})$$

"...so if the overall expression isn't fulfilled, then at least one of the clauses must not be satisfied."

"Let c be an unsatisfied clause," Miruka said, standing and scanning our faces. "What can you tell me about this clause c, Tetra?"

"Um, sorry, give me a minute to remember the definition of what a clause is."

"No need to apologize for that."

"So, a clause is some literals tied together with OR expressions, right?"

"How many?" Miruka promptly asked.

"How many literals? Uh..."

"Three," Lisa said, startling me—I had almost forgotten she was there.

"Three?" Tetra said. "Oh, right! Three! Because the logical expression we're talking about is called 3-CNF."

"Right. There's three literals, so the clause c looks like this."

$$\text{literal}_1 \lor \text{literal}_2 \lor \text{literal}_3$$

"Oh! Of course! Assignment a doesn't satisfy clause c, so literal_1 and literal_2 and literal_3 must all be false!"

"Very good," Miruka said. "We have three false literals in assignment a, and the true/false value of at least one of the variables among those in clause c must be incorrectly assigned."

"Interesting!" I said.

"Incorrectly assigned, how?" Tetra said.

"It's like this, Tetra," I said, glancing at Miruka. "Say clause c is $x_1 \lor \neg x_2 \lor \neg x_3$. If assignment a doesn't satisfy this clause, then in that assignment x_1 must be false, and x_2 and x_3 must be true. So to satisfy c, we need to flip the value of at least one of x_1, x_2, or x_3."

"Ah hah! But hang on, it seems like flipping just one of x_1, x_2, or x_3 should do the trick. Why do you say *at least one* of them needs to be changed?"

"Good point," I said. "You're right that we can satisfy c by flipping just one variable. But that change might have effects on other clauses, making them not satisfy the overall expression. That won't do—we need to make every clause work, and there's no easy way to know how many variables we have to flip to do so. Maybe it's just one, maybe it's more."

"Um, so..."

"It's also hard to know which ones to flip," I continued. "For example, say the logical expression looks like this."

$$\underbrace{(x_1 \lor x_2 \lor x_3)}_{\text{clause}_1} \land \cdots \land \underbrace{(\neg x_1 \lor x_2 \lor x_3)}_{\text{clause}_{123}}$$

"In this case, an assignment like this..."

$$(x_1, x_2, x_3) = (\text{False}, \text{False}, \text{False})$$

"... won't satisfy the expression. So what happens if we flip the value of x_1?"

$$(x_1, x_2, x_3) = (\underline{\text{True}}, \text{False}, \text{False})$$

"Now clause$_1$ is satisfied, but clause$_{123}$, which was okay before, isn't."

"Knock one down, another one pops up," Tetra said. "Kinda like whack-a-mole!"

9.4.3 Quantitative estimations

"Okay, I think I'm getting a good feel for random walks," Tetra said, "and after talking through all this I'm more comfortable with 3-CNF, clauses, and literals. But I'm still not sure how a randomized algorithm like this RANDOM-WALK-3-SAT can be faster than a brute-force algorithm."

"To answer that question," Miruka said, "we need a quantitative estimation with some clear preconditions. Luckily, we've just gotten a good clue as to where we should begin."

"What clue?"

"Well, like he just said, the true/false value of at least one of the three variables appearing in the unsatisfied clause c must be wrong."

"So we need to flip one, right?"

"But more specifically, the probability of picking a variable that needs to be changed from among the three in clause c is at least $\frac{1}{3}$."

"Oh, of course!" I said, louder than intended.

"That's the clue," Miruka said. "We randomly select a variable from among those in a non-satisfying clause and flip its value. By doing so, there's at least a $\frac{1}{3}$ probability that we've moved one step closer to finding an assignment that satisfies the logical expression."

"Got it," Tetra said, but then grimaced. "Or not. You said we might get one step closer to an assignment that works, but how can we know if the variable we selected really does take us there? I mean, we don't know, right? If we did, the problem would be solved. So it seems like no matter how many random walks we take, we can never know if we're getting closer to the right answer, or farther away."

Tetra's impassioned outburst stirred something within me. I felt like there was something I should be noticing, but I wasn't sure what. *What is it...?*

Miruka continued, softly. "You're right. We cannot know if we're moving closer or farther from the correct assignment. But there are some things we do know."

- The probability of moving closer after making one random selection is at least $\frac{1}{3}$.

- The probability of moving farther after making one random selection is at most $\frac{2}{3}$.

"So now that we've— What's wrong?"

Miruka was looking at me. Tetra and Lisa followed her gaze. I was rapidly shaking my head.

"I think I see it. The probability of getting closer, and the probability of getting farther."

"You want to say that—"

"Hold up, Miruka," I said. "There's another random walk here. A hidden one-dimensional random walk!"

"That's right," Miruka said, her cool contrasting with my excitement. "A random walk on the Hamming distance."

Somehow, I felt like I had been led to this.

9.4.4 Another Random Walk

"We want to know whether we're getting closer to the correct assignment, or farther away," Miruka dispassionately stated. "Let's bring some form to that concept, by adding 'distance' to the set of assignments."

"Distance?" Tetra said.

"Compare two assignments, a and b. Let's say that x_1 is true in a, but false in b. The values are mismatched. We'll also assume that x_2 is false in both a and b. Good so far, Tetra?"

"I'm good. We're comparing two assignments."

"When two assignments have many mismatches, we'll say the distance between them is large. When there are few mismatches, the distance between them is small. That's not much of a stretch, right? To quantitatively discuss distance, we'll define it as the number of mismatching variables between two assignments. A distance like that is called a Hamming distance."

"Hamming distance," Tetra whispered, jotting the term down in her notebook.

"For example, there are three disagreeing variables in these two assignments a and b, so their distance is 3."

Assignment a $\quad (x_1, x_2, x_3, x_4) = (\underline{\text{True}}, \text{True}, \underline{\text{False}}, \underline{\text{False}})$

Assignment b $\quad (x_1, x_2, x_3, x_4) = (\underline{\text{False}}, \text{True}, \underline{\text{True}}, \underline{\text{True}})$

"Variables x_1, x_3, and x_4 are mismatched. Right," Tetra said.

"Let's assume that some assignment satisfying a given logical expression exists. We'll call that assignment a^*. If more than one exists, we'll let one of them be a^*."

"Okay, a^* is the correct assignment," Tetra said. "Well, one of them, at least."

"So tell me, when will the distance between a and a^* be 0?" Miruka asked.

"The distance will be 0 when...uh...when they're the same!"

"Correct. And when that happens, assignment a satisfies the logical statement."

"Right!"

"Okay, next question. When will the distance between a and a^* be 1?"

"When variable x_1 in assignment a is mismatched!"

"Wrong."

"Oh, wait... No, it's when only one variable in a is mismatched!"

"Good. And if we flip the value of the mismatched variable, the assignment will satisfy the logical expression."

"Understood."

"Another question. By how much does the distance change when we flip the value of a single variable?"

"When we change a true to false or vice versa, right? If we flip a variable then we're either making a matching variable mismatched or a mismatched variable match, so the distance should either increase or decrease by 1."

"Well done. The assignment changes with each flip, getting either one step closer or one step farther from a^*."

9.4.5 Focusing on Rounds

"Okay, Miruka," Tetra said. "I think I'm good with distance. What's next?"

"We want to perform a quantitative analysis of the randomized algorithm RANDOM-WALK-3-SAT, and show that it will complete a search in less time than the 2^n required for a brute force search."

"Are we going to show that it has an order of $n \log n$?"

"That would be quite a feat. Actually, it would even be too much to hope for n^K. No, what we're hoping to do is reduce the base of 2^n—in other words, the 2—to some lower value."

"Oh, I see," Tetra said as she wrote in her notebook. "But you also said we're going to perform a quantitative analysis of the randomized algorithm. How do we do that?"

"Let's focus on one round, one of the rounds of random walks in which we take $3n$ steps."

"The inner while loop in the procedure, right?"

⋮

W5:	$k \leftarrow 1$
W6:	while $k \leqslant 3n$ do
W7:	if ⟨assignment a satisfies f⟩ then
W8:	return ⟨Satisfiable⟩
W9:	end-if
W10:	$c \leftarrow$ ⟨A clause in f not satisfied by assignment a⟩
W11:	$x \leftarrow$ ⟨Randomly select a variable from clause c⟩
W12:	$a \leftarrow$ ⟨Obtain an assignment by reversing the value of variable x in assignment a⟩
W13:	$k \leftarrow k + 1$
W14:	end-while

⋮

One round of a random walk of $3n$ steps.

"One round lasts for up to $3n$ steps. So what would be the minimal probability that at one of those steps we stumble upon a correct assignment a^*? That's what we want to know, the probability of round success. In other words, the probability that one round will

complete and output ⟨Satisfiable⟩. To estimate that from below—
Yes, Tetra?"

I looked to see Tetra with her hand raised, something she rarely did in the middle of an explanation.

"Um, doesn't it kind of feel like we've gotten off-topic? I mean, we want to estimate the number of execution steps, so why are we talking about probabilities?"

"Hmph. Okay, let's talk about that first. Here's what the outer loop looks like."

$$
\begin{aligned}
&\vdots \\
W2:\quad &r \leftarrow 1 \\
W3:\quad &\text{while } r \leqslant R \text{ do} \\
W4:\quad &\quad a \leftarrow \langle \text{Randomly assign } n \text{ variables} \rangle \\
&\vdots \\
&\quad \text{(One round of a random walk of } 3n \text{ steps)} \\
&\vdots \\
W15:\quad &\quad r \leftarrow r + 1 \\
W16:\quad &\text{end-while} \\
&\vdots
\end{aligned}
$$

"If we find an assignment that satisfies the logical expression, the algorithm ends there, so a higher probability of round success means fewer outer loops executed. That means by estimating the probability of round success, we're also estimating the number of execution steps we can expect."

"I see. That makes sense." Reviewing her notes, Tetra continued. "One more question, though maybe not quite on topic... An algorithm is supposed to output a correct answer, right? So why does RANDOM-WALK-3-SAT output ⟨*Probably* not satisfiable⟩?"

"The correctness of a randomized algorithm is related to probability," Miruka said. "If RANDOM-WALK-3-SAT outputs ⟨Satisfiable⟩, then the expression is definitely satisfiable. But the expression may also be satisfiable if it outputs ⟨Likely unsatisfiable⟩.

A solution may have been overlooked. So again, estimating probabilities is very important."

"Estimating the probability of overlooking a satisfying assignment?"

"Right. A randomized algorithm like RANDOM-WALK-3-SAT—where one of two outputs is 100% correct and the other is correct with some probability—is called a Monte Carlo algorithm with one-sided error. In RANDOM-WALK-3-SAT, if you want to increase the probability that an output of ⟨Likely unsatisfiable⟩ will only occur when there really isn't a satisfying assignment, you just need to increase the number of rounds R. Of course the trade-off is that you might have to run more execution steps, so deciding how large to set R is a judgment call."

"You were saying something about estimating the probability of round success?" I said.

"Right. Say that if we use some constant M greater than 1, we can estimate from below by saying that the probability of round success is at least $\frac{1}{M^n}$. Then we set the number of rounds R as $R = K \cdot M^n$. In other words, the reciprocal of the probability of round success multiplied by K. We can freely choose K as some number that doesn't rely on n."

"What do we do next?" Tetra asked.

"Next we estimate from above, we find the probability that RANDOM-WALK-3-SAT overlooks some assignment that satisfies the logical expression. This 'probability of overlooking' equals the Rth power of the probability of a round failure. We can estimate the probability of round failure as at most $1 - \frac{1}{M^n}$, so we can estimate from above like this."

Probability of overlooking = (Probability of round failure)^R

$$\leqslant \left(1 - \frac{1}{M^n}\right)^R$$

$$= \left(1 - \frac{1}{M^n}\right)^{K \cdot M^n}$$

$$\leqslant e^{-\frac{1}{M^n} \cdot K \cdot M^n}$$

$$= e^{-K}$$

$$= \frac{1}{e^K}$$

"We can choose any value for K that we like, so we can set some upper limit $\frac{1}{e^K}$, and say that the probability of an overlook is at most that."

$$\text{Probability of overlooking} \leqslant \frac{1}{e^K}$$

"Having come this far, exponential growth is on our side. By making K a little larger, we can make $\frac{1}{e^K}$ way smaller. In other words, we can constrain overlooks to a very low probability. When we do that, the exponential with the number of rounds R is of order M^n, no matter the value of the constant K."

"I see. So we can estimate the probability of round success as at least $\frac{1}{M^n}$. We want to be able to say that's the minimum probability that we will find within one round an assignment of variables that satisfies the logical expression."

"I'm sorry, one question," Tetra said. "About this part."

$$\left(1 - \frac{1}{M^n}\right)^{K \cdot M^n} \leqslant e^{-\frac{1}{M^n} \cdot K \cdot M^n}$$

"How do we know that this is true?"
"Look at this graph of $y = e^x$," Miruka said.

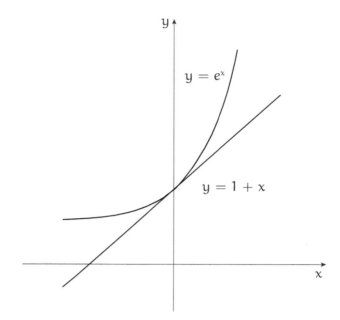

"See how $1 + x \leqslant e^x$ will hold for any value of x? Now just set $x = -\frac{1}{M^n}$."

$$1 - \frac{1}{M^n} \leqslant e^{-\frac{1}{M^n}}$$

"Now raise both sides to the $K \cdot M^n$ power. Both sides are positive, so the direction of the inequality sign won't change."

$$\left(1 - \frac{1}{M^n}\right)^{K \cdot M^n} \leqslant e^{-\frac{1}{M^n} \cdot K \cdot M^n}$$

"Before, we just did all this in one step," Miruka said.

"The library is *closed*," came Ms. Mizutani's announcement, and thus ended another fun-filled day of math in the school library.

9.5 At Home

9.5.1 Estimating Luck

That night I was studying alone in my room. I really enjoyed doing math with Miruka and Tetra, but time to work on my own was also important.

My inner monologue was about the randomized algorithm we had been discussing.

$\vdots$

$W5$:	$k \leftarrow 1$
$W6$:	while $k \leqslant 3n$ do
$W7$:	if ⟨assignment a satisfies f⟩ then
$W8$:	return ⟨Satisfiable⟩
$W9$:	end-if
$W10$:	$c \leftarrow$ ⟨A clause in f not satisfied by assignment a⟩
$W11$:	$x \leftarrow$ ⟨Randomly select a variable from clause c⟩
$W12$:	$a \leftarrow$ ⟨Obtain an assignment by reversing the value of variable x in assignment a⟩
$W13$:	$k \leftarrow k + 1$
$W14$:	end-while

$\vdots$

One round of a random walk of $3n$ steps.

I jotted down a couple of notes to confirm where I was, and where I wanted to go.

Question: What do I want to know?

Answer: The round success probability. Namely, I want to estimate from below the probability of ⟨Satisfiable⟩ being output within $3n$ steps of a random walk.

Question: What do I know so far?

Answer: By flipping the true/false value of one variable in a randomly selected non-satisfying clause c, the probability of getting closer to a correct assignment a^* is at least $\frac{1}{3}$.

I noticed one thing right off: *This is another coin-flipping problem!*

· The probability of flipping heads is at least $\frac{1}{3}$.

· The probability of flipping tails is at most $\frac{2}{3}$.

I wanted to repeat coin flips like this to describe a random walk in which a heads took me one step closer to my destination, and a tails took me one step farther away. One problem, though: I didn't know the distance—in other words, the number of mismatched variables— between the current assignment a and a correct assignment a^*. Assignment a is randomly determined when the round begins, so there was no way to know the distance between a and a^*.

But hang on! I can know the probability!

I used $p(m)$ to describe the probability that the distance from a^* when starting the round was m, like this.

$$p(m) = \frac{\text{probability that the distance between } a^* \text{ and}}{\text{a randomly selected assignment a is } m}$$

So now the question was, can I find $p(m)$?

Actually, maybe that's not so hard...

Given n variables, randomly determining an assignment means randomly selecting n true-or-false values. There are 2^n ways to do that. The number of cases in which m of n variables are mismatched is the same as the number of ways to select m items from n, or $\binom{n}{m}$. Each combination could occur with equal probability, so the probability would be this.

$$p(m) = \frac{\text{cases where } m \text{ of } n \text{ variables are mismatched}}{\text{all possible cases}} = \frac{\binom{n}{m}}{2^n} = \frac{1}{2^n}\binom{n}{m}$$

Now we're getting somewhere.

My goal was to estimate the probability of round success from below, from smaller values. I wanted to find the absolute minimal probability of success.

Okay, so just how lucky can I get?

Letting m be the initial number of differences, what would be the luckiest of all possible assignments? Well that's easy—the luckiest case is the one where *every* selection of a variable to flip is one that is mismatched from an appropriate assignment.

If m is the number of mismatches in the initial assignment, then m is also the distance. Then in this luckiest of situations, over m continuous steps we would step closer to a^*, bringing the distance to

0 and giving a satisfying assignment. The luckiest of random walks would be one that heads directly to the goal.

Okay, so let's calculate that.

Let q(m) be the probability of hitting this luckiest of cases, in other words the probability of decreasing the distance m times in a row. There's at least a 1/3 chance of decreasing the distance with each step, so—

$$q(m) \geq \left(\frac{1}{3}\right)^m$$

From that, I got this.

p(m)	=	probability of starting from distance m	= $\frac{1}{2^n}\binom{n}{m}$
q(m)	=	probability of m continuous distance reductions	$\geq \left(\frac{1}{3}\right)^m$

This left me wondering how I could handle m, so I decided to think of things a little more concretely. In the case where assignment a satisfied the logical expression f, m would be 0. Or m could be 1, in the case where there was just one mismatched variable. Similarly, m could be 2, or 3, or anything up to n.

So m can take any value, $0, 1, 2, \ldots, n$, and there would be no leaks and no dupes.

That's it! Got it!

There would be n + 1 possibilities for m when we start a round, namely $m = 0, 1, 2, 3, \ldots, n$, and that accounted for all possibilities. Further, each possible condition excluded all others. In other words, I could just calculate the probability of round success for each value of m, and add all those probabilities together. Doing that would let me find the round success probability in the luckiest case, regardless of the m value.

The round success probability when m is determined would be the product of two values: the probability of the initial distance being m, times the probability of m continuous lucky variable flips. In other words, the round success probability in the luckiest of all cases would be

$$p(m)q(m),$$

and I wanted to add these up for $m = 0, 1, 2, 3, \ldots, n$.

Round success probability

$\geqslant$ The round success probability in the luckiest case

$= \underbrace{p(0)q(0)}_{m=0 \text{ case}} + \underbrace{p(1)q(1)}_{m=1 \text{ case}} + \underbrace{p(2)q(2)}_{m=2 \text{ case}} + \underbrace{p(3)q(3)}_{m=3 \text{ case}} + \cdots + \underbrace{p(n)q(n)}_{m=n \text{ case}}$

$= \sum_{m=0}^{n} p(m)q(m)$

$\geqslant \sum_{m=0}^{n} \frac{1}{2^n} \binom{n}{m} \left(\frac{1}{3}\right)^m \qquad$ from $p(m) = \frac{1}{2^n}\binom{n}{m}$ and $q(m) \geqslant \left(\frac{1}{3}\right)^m$

There it is! I have a good estimation from below!

Round success probability $\geqslant \sum_{m=0}^{n} \frac{1}{2^n} \binom{n}{m} \left(\frac{1}{3}\right)^m$

It wasn't pretty, though. If possible, I wanted to clean up this inequality a bit.

Problem 9-2 (Simplifying a sum)

Simplify the following sum:

$$\sum_{m=0}^{n} \frac{1}{2^n} \binom{n}{m} \left(\frac{1}{3}\right)^m.$$

9.5.2 Simplifying the Sum

Talking about logic was fun. Talking about random walks was fun. But funnest of all was doing what still felt to me like "real math." When I had everything in the form of mathematical notation, I felt like I really knew who my opponent was.

Now stepping into the ring was

$$\sum_{m=0}^{n} \frac{1}{2^n} \binom{n}{m} \left(\frac{1}{3}\right)^m,$$

and I wanted to see if I could cut him down to size.

First off, get the mechanical stuff out of the way. There were no m's in $\frac{1}{2^n}$, so I could move that outside of the sigma.

$$\sum_{m=0}^{n} \frac{1}{2^n} \binom{n}{m} \left(\frac{1}{3}\right)^m = \frac{1}{2^n} \sum_{m=0}^{n} \binom{n}{m} \left(\frac{1}{3}\right)^m$$

Next, multiply $\binom{n}{m}$ and $\left(\frac{1}{3}\right)^m$, and add up the various m's—a sum of products.

This reminded me of Tetra, who had so much fun finding sums of products. Tetra, so direct in how she studied and how she spoke. Tetra, who so diligently studied what she was taught. Tetra, who still got uneasy when faced with lots of variables, like in the binomial theorem...

The binomial theorem?

$$(x+y)^n = \sum_{m=0}^{n} \binom{n}{m} x^{n-m} y^m \quad \text{(the binomial theorem)}$$

The binomial theorem! That's it! The binomial theorem with $x = 1, y = \frac{1}{3}$ is the key to simplifying the sum!

$$\frac{1}{2^n} \sum_{m=0}^{n} \binom{n}{m} \left(\frac{1}{3}\right)^m = \frac{1}{2^n} \sum_{m=0}^{n} \binom{n}{m} \cdot 1^{n-m} \cdot \left(\frac{1}{3}\right)^m$$

$$= \frac{1}{2^n} \left(1 + \frac{1}{3}\right)^n \quad \text{(binomial theorem)}$$

$$= \left(\frac{1}{2}\right)^n \left(\frac{4}{3}\right)^n$$

$$= \left(\frac{1}{2} \cdot \frac{4}{3}\right)^n$$

$$= \left(\frac{2}{3}\right)^n$$

That's it! And it's even simpler than I imagined!

Answer 9-2 (Simplifying a sum)

$$\sum_{m=0}^{n} \frac{1}{2^n} \binom{n}{m} \left(\frac{1}{3}\right)^m = \left(\frac{2}{3}\right)^n$$

I now knew how to estimate the round success probability when using n variables.

$$\text{Round success probability} \geq \left(\frac{2}{3}\right)^n$$

9.5.3 Progress

I was on a roll. I had found the round success probability to be at least $\left(\frac{2}{3}\right)^n$. Putting it into the "at least $\frac{1}{M^n}$" form that Miruka had used, M would be the reciprocal of $\frac{2}{3}$, in other words $\frac{3}{2}$. So the exponential function part of the round count $R = K \cdot M^n$ would be this.

$$M^n = \left(\frac{3}{2}\right)^n = 1.5^n$$

In the case of a brute force the 2^n repetitions had a base of 2. I had just estimated the exponential function part of the round count in RANDOM-WALK-3-SAT as 1.5^n, which gave a base of 1.5. So I had gone from 2.0 to 1.5. Progress!

9.6 IN THE LIBRARY

9.6.1 Independence and Exclusion

The next day in the library I presented what I had found to the usual gang—Tetra, Miruka, and the ever-silent Lisa.

Tetra gave me a round of applause. "So 2^n became 1.5^n!"

I tried to restrain my excitement as I answered. "Yeah, which reduced the base from 2.0 to 1.5." I was quite pleased at having managed an estimation without reading the paper Miruka had mentioned. "It all fell out from noticing that I could use the binomial theorem."

"An interesting approach," Miruka said. "Of course, unlike when searching by brute force there's no guarantee you'll find a satisfying assignment at order 1.5^n. Even so, when faced with an exponential order like 2^n, it's important to at least try to reduce the base as much as possible. Randomized algorithms are one important tool for doing so."

"I'm starting to really like this coin-flipping idea," Tetra said. "Viewing things like a 3n-step random walk or the R rounds as coin flips really makes things easy to understand.

"And each coin flip is independent."

"Independent how?"

"Two events A and B are independent if they exert no influence on each other. That means the probability of both A and B occurring equals the product of Pr(A) and Pr(B)."

$$\Pr(A \text{ and } B) = \Pr(A) \times \Pr(B) \qquad (\text{when } A, B \text{ are independent})$$

Tetra jotted this down in her notebook. "Independent... Got it. This is different from 'exclusive,' right?"

Miruka nodded. "It is. If two events A and B are exclusive, then the probability of A or B occurring equals the sum of Pr(A) and Pr(B)."

$$\Pr(A \text{ or } B) = \Pr(A) + \Pr(B) \qquad (\text{when } A, B \text{ are exclusive})$$

9.6.2 Precise Estimates

"So let's discuss the estimation I read about in that paper," Miruka said.

Tetra's eyebrows rose. "What? The answer wasn't 1.5?"

"It's a little bit more precise."

I slapped my forehead. "Of course! Stirling's approximation!"

"We'll use Stirling's approximation, sure," Miruka said. "But first we'll use another familiar friend—your generalized solution to the piano problem[1]."

General solution to the piano problem

Connecting $a+b$ adjacent notes without going below the starting note, the number of possible melodies ending $a-b-1$ notes above the starting note is

$$\frac{a-b}{a+b} \cdot \binom{a+b}{a}.$$

---

[1] Called "Bertrand's ballot theorem" in probability theory.

"So let's make a more precise estimation of RANDOM-WALK-3-SAT," Miruka said. "Say that the assignment given at the start of our random walk is m steps from a satisfying assignment a^*. We start our random walk from there, and say that the logical expression is satisfied if the distance becomes 0. When you estimated the probability of round success last night, you only considered the luckiest case, where you started m steps from a^*, and the distance became 0 after advancing just m steps. But in most cases, you'll have to wander around a bit before the distance becomes 0. When you consider all the possible wanderings the number of routes to a^* will increase, which will increase the probability of round success, and thereby decrease the order of the round count. Good so far?"

I nodded. "Sure, makes sense."

"Okay, so here's what we want to think about. Somewhere along the way while trying to reduce the distance from m to 0, we will take i steps that move us away from a^*. To get those i steps back, at some point we will have to get i steps closer. Combining that with the m steps needed to go from m to 0, the total steps from our starting position to a^* will be $m + 2i$. We can ignore any routes where the number of distancing steps i exceeds m, since we can consider any such route as unable to reach a^*.

"Compare these two graphs, one showing how the distance to a^* changes as we proceed, the other showing how the pitches change in the piano problem. See how there's a one-to-one correspondence between a random walk and a piano melody? One is a left–right reflection of the other."

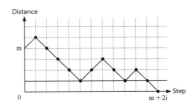
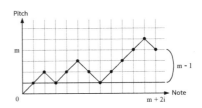

"In the general solution to the piano problem, we wanted to end $a - b - 1$ notes above the starting note, and we found that we can get the number of melodies comprising $a + b$ notes like this."

$$\frac{a-b}{a+b}\binom{a+b}{a}$$

"If we consider this as a random walk, we get this system of equations."

$$\begin{cases} a - b - 1 = m - 1 & \text{(how much higher to go)} \\ a + b = m + 2i & \text{(number of notes to use)} \end{cases}$$

"Solving this we get $(a, b) = (m + i, i)$, and we can find the number of paths for a random walk."

$$\frac{a-b}{a+b}\binom{a+b}{a} = \frac{a-b}{a+b}\binom{a+b}{b} \quad \text{because } \binom{a+b}{a} = \binom{a+b}{b}$$
$$= \frac{m}{m+2i}\binom{m+2i}{i}$$

"With this, we have the number of paths starting from a distance m, arriving at a^* after getting i steps farther from it. Next, we need to calculate the probability of round success.

"We'll let $P(m, i)$ be the probability of arriving at a^* from distance m, with i steps of detour, and estimate from below. We're going to make $m + 2i$ coin flips, with heads meaning we go farther away, and tails meaning we get closer. In other words, we're getting i heads. There was a $\frac{1}{3}$ probability of getting tails in the worst-case scenario, which means a $\frac{2}{3}$ probability of heads. So $P(m, i)$ looks like this."

$$P(m, i) \geqslant \underbrace{\frac{m}{m+2i}\binom{m+2i}{i}}_{\text{paths}} \underbrace{\left(\frac{2}{3}\right)^i}_{\text{distancing part}} \underbrace{\left(\frac{1}{3}\right)^{m+i}}_{\text{nearing part}}$$

"In this estimation, we're ignoring all paths where the number of distancing steps i is bigger than m. We've estimated $P(m, i)$ from below, so we'll use $Q(m)$ to represent the probability of arriving at a^* from distance m, regardless of the number of distancing steps.

This will be within the range of valid values for i, namely $0 \leqslant i \leqslant m$, so we can find $Q(m)$ as a sum of $P(m,i)$'s."

$$Q(m) = \sum_{i=0}^{m} P(m,i)$$

$$\geqslant \sum_{i=0}^{m} \underbrace{\frac{m}{m+2i}}_{\alpha} \binom{m+2i}{i} \underbrace{\left(\frac{2}{3}\right)^{i}}_{\beta} \underbrace{\left(\frac{1}{3}\right)^{m+i}}_{\gamma}$$

"We can estimate α, β, γ from below, using the fact that m is the maximum value for i."

$$\begin{cases} \alpha : & \dfrac{m}{m+2i} \geqslant \dfrac{m}{m+2m} = \dfrac{1}{3} \\ \beta : & \left(\dfrac{2}{3}\right)^{i} \geqslant \left(\dfrac{2}{3}\right)^{m} \\ \gamma : & \left(\dfrac{1}{3}\right)^{m+i} \geqslant \left(\dfrac{1}{3}\right)^{m+m} = \left(\dfrac{1}{3}\right)^{2m} \end{cases}$$

"We'll use this to proceed with our estimation of $Q(m)$."

$$Q(m) \geqslant \sum_{i=0}^{m} \underbrace{\frac{m}{m+2i}}_{\alpha} \binom{m+2i}{i} \underbrace{\left(\frac{2}{3}\right)^{i}}_{\beta} \underbrace{\left(\frac{1}{3}\right)^{m+i}}_{\gamma}$$

$$\geqslant \sum_{i=0}^{m} \frac{1}{3} \binom{m+2i}{i} \left(\frac{2}{3}\right)^{m} \left(\frac{1}{3}\right)^{2m} \qquad \text{use } m \text{ as the max. value for } i$$

$$= \frac{1}{3} \left(\frac{2}{3}\right)^{m} \left(\frac{1}{3}\right)^{2m} \sum_{i=0}^{m} \binom{m+2i}{i} \qquad \text{move parts with no } i \text{ outside}$$

$$= \frac{1}{3} \left(\frac{2}{27}\right)^{m} \sum_{i=0}^{m} \binom{m+2i}{i}$$

"Taking out one term from the sum, we can create an inequality..."

$$\sum_{i=0}^{m} \binom{m+2i}{i} \geqslant \binom{m+2m}{m} = \binom{3m}{m}$$

"...and using this we can continue with our estimation of Q(m)."

$$Q(m) \geqslant \frac{1}{3}\left(\frac{2}{27}\right)^m \underbrace{\sum_{i=0}^{m} \binom{m+2i}{i}}$$

$$\geqslant \underbrace{\frac{1}{3}}_{\text{constant}} \underbrace{\left(\frac{2}{27}\right)^m}_{\text{power}} \binom{3m}{m}$$

"Let's estimate $\binom{3m}{m}$ as a power. To do that, we'll use a new power tool, Stirling's approximation."

9.6.3 Stirling's Approximation

Miruka continued filling my notebook with equations as she went on with her lecture.

"Stirling's approximation is a good way of estimating values for $n!$."

Stirling's approximation

For sufficiently large n, $n!$ can be approximated as $\sqrt{2\pi n}\left(\frac{n}{e}\right)^n$.

$$n! \sim \sqrt{2\pi n}\left(\frac{n}{e}\right)^n$$

Note: The limit as $n \to \infty$ of the ratio of the two sides equals 1. In other words,

$$\lim_{n\to\infty} \frac{n!}{\sqrt{2\pi n}\left(\frac{n}{e}\right)^n} = 1$$

"We're going to use these inequalities, which are related to Stirling's approximation."

$$n! \leqslant \sqrt{2\pi n}\left(\frac{n}{e}\right)^n e^{\frac{1}{12n}} \qquad \text{evaluation from above (U)}$$

$$n! \geqslant \sqrt{2\pi n}\left(\frac{n}{e}\right)^n \qquad \text{evaluation from below (L)}$$

"Okay, now let's use inequalities (U) and (L) to estimate $\binom{3m}{m}$ from below. From the definition, we can write $\binom{3m}{m}$ as factorials like

this."
$$\binom{3m}{m} = \frac{(3m)!}{(1m)!\,(2m)!}$$

"I wrote m as 1m here to make it easier to see the pattern. We want to estimate $\binom{3m}{m}$ from below, so we'll use a largish denominator and a smallish numerator. In other words, we'll use estimations from above (U) for (1m)! and (2m)! and estimation from below (L) for (3m)!."

$$(1m)! \leq \sqrt{2\pi \cdot 1m}\left(\frac{1m}{e}\right)^{1m} e^{\frac{1}{12 \cdot 1m}} \qquad \text{from (U)}$$

$$(2m)! \leq \sqrt{2\pi \cdot 2m}\left(\frac{2m}{e}\right)^{2m} e^{\frac{1}{12 \cdot 2m}} \qquad \text{from (U)}$$

$$(3m)! \geq \sqrt{2\pi \cdot 3m}\left(\frac{3m}{e}\right)^{3m} \qquad \text{from (L)}$$

"Here's an estimation of the denominator in $\frac{(3m)!}{(1m)!\,(2m)!}$."

$$(1m)!\,(2m)! \leq \sqrt{2\pi \cdot 1m}\left(\frac{1m}{e}\right)^{1m} e^{\frac{1}{12 \cdot 1m}} \cdot \sqrt{2\pi \cdot 2m}\left(\frac{2m}{e}\right)^{2m} e^{\frac{1}{12 \cdot 2m}}$$
$$= 2\pi \cdot \sqrt{2} \cdot m \cdot 4^m \cdot m^{3m} \cdot e^{-3m} \cdot e^{\frac{1}{12m} + \frac{1}{24m}}$$

"And here's an estimation of the numerator in $\frac{(3m)!}{(1m)!\,(2m)!}$."

$$(3m)! \geq \sqrt{2\pi \cdot 3m}\left(\frac{3m}{e}\right)^{3m}$$
$$= \sqrt{2\pi} \cdot \sqrt{3} \cdot \sqrt{m} \cdot 27^m \cdot m^{3m} \cdot e^{-3m}$$

"With this, we can estimate $\binom{3m}{m}$"

$$\binom{3m}{m} = \frac{(3m)!}{(1m)!\,(2m)!}$$
$$\geq \frac{\sqrt{2\pi} \cdot \sqrt{3} \cdot \sqrt{m} \cdot 27^m \cdot m^{3m} \cdot e^{-3m}}{2\pi \cdot \sqrt{2} \cdot m \cdot 4^m \cdot m^{3m} \cdot e^{-3m} \cdot e^{\frac{1}{12m} + \frac{1}{24m}}}$$
$$= \frac{\sqrt{3} \cdot 27^m}{\sqrt{2\pi} \cdot \sqrt{2} \cdot \sqrt{m} \cdot 4^m \cdot e^{\frac{1}{8m}}}$$
$$= \frac{\sqrt{3}}{2\sqrt{\pi}} \cdot e^{-\frac{1}{8m}} \cdot \frac{1}{\sqrt{m}} \cdot \left(\frac{27}{4}\right)^m$$

"When $m = 1, 2, 3, \ldots, n$, we use $e^{-\frac{1}{8m}} \geqslant e^{-\frac{1}{8}}$"

$$\geqslant \underbrace{\frac{\sqrt{3}}{2\sqrt{\pi}} \cdot e^{-\frac{1}{8}}}_{\text{constant}} \cdot \frac{1}{\sqrt{m}} \cdot \underbrace{\left(\frac{27}{4}\right)^m}_{\text{as a power}}$$

"Let's summarize the constants as C."

$$C = \frac{\sqrt{3}}{2\sqrt{\pi}} \cdot e^{-\frac{1}{8}}$$

"Now let's estimate $\binom{3m}{m}$."

$$\binom{3m}{m} \geqslant \frac{C}{\sqrt{m}} \left(\frac{27}{4}\right)^m$$

"Okay, back to estimating $Q(m)$."

$$Q(m) \geqslant \frac{1}{3}\left(\frac{2}{27}\right)^m \cdot \binom{3m}{m}$$

$$\geqslant \frac{1}{3}\left(\frac{2}{27}\right)^m \cdot \frac{C}{\sqrt{m}} \left(\frac{27}{4}\right)^m$$

$$= \frac{C}{3} \frac{1}{\sqrt{m}} \left(\frac{2}{27} \cdot \frac{27}{4}\right)^m$$

$$= \frac{C}{3} \frac{1}{\sqrt{m}} \left(\frac{1}{2}\right)^m$$

"Let $C' = \frac{C}{3}$."

$$= \frac{C'}{\sqrt{m}} \left(\frac{1}{2}\right)^m$$

"Now we can estimate the probability of round success from below."

Round success probability

$$= \sum_{m=0}^{n} \begin{matrix}\text{Probability of initial}\\ \text{distance being } m\end{matrix} \cdot Q(m)$$

$$= \sum_{m=0}^{n} \frac{1}{2^n} \binom{n}{m} \cdot Q(m)$$

$$\geq \sum_{m=0}^{n} \frac{1}{2^n} \binom{n}{m} \cdot \frac{C'}{\sqrt{m}} \left(\frac{1}{2}\right)^m$$

$$\geq \frac{C'}{\sqrt{n}} \frac{1}{2^n} \sum_{m=0}^{n} \binom{n}{m} \left(\frac{1}{2}\right)^m \qquad \text{from } \frac{1}{\sqrt{m}} \geq \frac{1}{\sqrt{n}}$$

$$= \frac{C'}{\sqrt{n}} \frac{1}{2^n} \underbrace{\sum_{m=0}^{n} \binom{n}{m} 1^{n-m} \left(\frac{1}{2}\right)^m}_{} \qquad \text{prepare for the binom. th.}$$

$$= \frac{C'}{\sqrt{n}} \frac{1}{2^n} \underbrace{\left(1 + \frac{1}{2}\right)^n}_{} \qquad \text{use the binomial theorem}$$

$$= \frac{C'}{\sqrt{n}} \left(\frac{1}{2} \cdot \frac{3}{2}\right)^n$$

$$= \frac{C'}{\sqrt{n}} \left(\frac{3}{4}\right)^n$$

"So now we've estimated the probability of round success, like this."

$$\text{Round success probability} \geq \frac{C'}{\sqrt{n}} \left(\frac{3}{4}\right)^n$$

"Taking the reciprocal of the right side, we get this..."

$$\frac{\sqrt{n}}{C'} \left(\frac{4}{3}\right)^n$$

"...so we can estimate the exponential part of the round count like this."

$$\text{Exponential part of the round count} \leq \left(\frac{4}{3}\right)^n = (1.333\cdots)^n < 1.334^n$$

"So in the end, we've estimated the exponential part of the number of rounds to be at most 1.334^n. That's quite an improvement over your 1.5^n." Miruka put down my pencil and sat back in her seat. "Done and done. Reading math papers can be lots of fun."

"Stirling's approximation, huh," I said.

A roadmap for estimating the base in the exponent part of the round count

Random walk start assumes a distance of
m steps from a satisfying assigment a^*.
Find the number of paths using the piano
problem, and estimate probabilities as coin flips.

$\downarrow$ estimate $P(m,i)$ as the probability of reaching a^* from distance m and detour of i steps.

$$P(m,i) \geqslant \underbrace{\frac{m}{m+2i}\binom{m+2i}{i}}_{\text{no. of routes}} \underbrace{\left(\frac{2}{3}\right)^i}_{\text{distancing}} \underbrace{\left(\frac{1}{3}\right)^{m+i}}_{\text{approaching}}$$

$\downarrow$ estimate $Q(m)$ as the probability of reaching a^* from distance m

$$Q(m) = \sum_{i=0}^{m} P(m,i) \geqslant \frac{1}{3}\left(\frac{2}{27}\right)^m \binom{3m}{m}$$

$\downarrow$ estimate $\binom{3m}{m}$ using Stirling's approximation

$$\binom{3m}{m} = \frac{(3m)!}{(1m)!\,(2m)!} \geqslant \frac{C}{\sqrt{m}}\left(\frac{27}{4}\right)^m$$

$\downarrow$ estimate round success probability using the binomial theorem

$$\text{Round success probability} = \sum_{m=0}^{n} \frac{1}{2^n}\binom{n}{m} \cdot Q(m) \geqslant \frac{C'}{\sqrt{n}}\left(\frac{3}{4}\right)^n$$

$\downarrow$ estimate exponential part of round count using reciprocal

$$\text{Exponential part of round count} \leqslant \left(\frac{4}{3}\right)^n < 1.334^n$$

Tetra sighed and slowly spoke. "I really need more tools in my toolbox. All kinds of tools, not just more skill at manipulating equations. Estimating the magnitudes of factors, using Stirling's approximation to estimate numbers of combinations, estimations from above and below, estimating the probability of round success, estimating the probability of failure, estimations of the probability of overlooking solutions in randomized algorithms... I need so much more knowledge about numbers and equations, not to mention a better sense of what to do with them. And the endurance to do it all!"

"Don't give up," I said. "You'll get there. Oops, looks like it's almost time for Ms. Mizutani to make her appearance."

Tetra gave a mischievous grin. "What do you say we mix things up today?"

"Mix things up how?" Miruka asked.

Tetra gestured for a huddle, and explained.

Ms. Mizutani marched out of her office right on schedule, wearing her usual tight skirt and dark shades. She followed her usual route to the center of the library to make her announcement, and just before she reached it...

"The library is *closed!*" we all shouted. Even Lisa joined in, though much more quietly.

Ms. Mizutani showed no reaction.

"The library is *closed!*" she announced.

9.7 ON THE WAY HOME

9.7.1 Olympics

We wound through the narrow streets toward the train station, the same path as always. I wondered what would happen if we instead took random walks toward our goal. Might we discover some interesting new place?

Tetra, reviewing her notes as we walked, broke the silence. "By the way, why does the algorithm we looked at run the inner loop $3n$ times? What's so special about $3n$?"

Miruka replied, "In our estimation we considered the case where the random walk goes i steps farther from its goal, in which case the number of steps became $m + 2i$. The maximum value of i is m, and the maximum value of m is n, so in order to arrive at a^*

within a round in which you've detoured by i steps, you need at least $m + 2i \leqslant n + 2n = 3n$ loops."

"Oh, I see! One more question about that inner loop, though. It seems like we paid the most attention to the number of times we executed the outer loop, but wouldn't all that stuff that was written out as words also take up a lot of time? Like, the stuff you have to do here where it says ⟨ Randomly assign n variables ⟩."

"That does take time, but procedures like that will have only polynomial orders. We assume so, at least. The line you mentioned is just storing values in n variables, which is of order $O(n)$, and the order for checks for the line ⟨ A clause in f is not satisfied by assignment a ⟩ will just be the number of clauses. If we don't allow repetition over any given clause, we can restrain the order of the number of clauses to some polynomial multiple of the number of variables. Again, no problem."

"Huh. Kinda reminds me of the Olympics."

"What does?" I asked.

"The 3-SAT algorithm. It's like a competition to see how low you can make the order, sort of like how everyone tries to set a new world record for the hundred-meter dash. Doesn't that feel like an Olympic event?"

"More like everyone competing to see who can be first to prove Fermat's last theorem."

"Oh, that's a good one too."

"Dream up an algorithm with a smaller order, analyze it, and write a paper," Miruka said. "Other mathematicians will read about your method, improve on it, and write their own paper. That's how we advance human knowledge. The paper I read was called 'A probabilistic algorithm for k-SAT and constraint satisfaction problems,' written in 1999 by Uwe Schöning[2]. When this paper was published, the 'world record' for 3-SAT became 1.334^n."

"Was it written in English?" Tetra asked.

"Of course. English is the lingua franca of academia."

"And most important of all, when you write a paper, accurately pass on something of value, right?"

---

[2] Ref. [28]

9.8 AT HOME

9.8.1 Logic

"Satisfiability problems, huh?" Yuri said.

The usual weekend, the usual room, the usual conversation with Yuri.

Or not? Something's different about her...

"Hey, Yuri. What's with the ribbon?" I asked.

"I can't believe you noticed."

"Of course I did. It's pretty. Looks good on you."

Yuri grinned. "Why thanks. But anyway, researching how hard a problem is sounds like fun. Not just solving a problem, but finding algorithms that can solve it? Problems about problems! Cool!"

"The important thing is that it allows quantitative estimations, using inequalities."

"Huh. Interesting that you can transform logic into inequalities."

"All math is connected somewhere, somehow."

"Even in computers."

"Apparently so. Lisa's a big help in that respect."

Yuri scowled. "Who's this Lisa chick?"

"A first-year student at our school. She's a whiz with computers."

Yuri paused to think a moment. "Does she know the Fibonacci sign?"

"Sure. She does it in binary, though. She's Dr. Narabikura's daughter. Speaking of which, there's going to be a conference at the Narabikura Library."

I found the pamphlet and showed it to Yuri. I explained what conferences are all about, and told her how this one was about computer science, and that there would be a session for junior high students.

She read the description of that session. "Oh, Miruka's going to be teaching it?"

"She was going to, but it looks like Tetra will be filling in for her."

"Aww, too bad. But it's for students my age? That means..."

"That means what?"

"Nah, nothing. Never mind. Still, it's hard to imagine Tetra giving a lecture. I'll bet she trips over the podium."

"I'm sure she won't. Well, pretty sure."

> The most famous unsolved problem in all of
> computer science is to find an efficient way to decide
> whether a given Boolean function is satisfiable or
> unsatisfiable.... When you hear about this problem
> for the first time, you might be tempted to ask a
> question of your own in return: "What? Are you
> serious that computer scientists still haven't figured
> out how to do such a simple thing?"
>
> DONALD KNUTH [22]

CHAPTER **10**

Randomized Algorithms

> I only observe this in particular, to show the reason why so much of my time went away with so little work—viz. that what might be a little to be done with help and tools, was a vast labour and required a prodigious time to do alone, and by hand. But notwithstanding this, with patience and labour I got through everything that my circumstances made necessary to me to do, as will appear by what follows.
>
> Daniel Defoe
> *Robinson Crusoe*

10.1 At the Restaurant

10.1.1 Rain

"I'm so very sorry!" Tetra said.

"Don't worry about it," I replied. Lisa remained silent.

We were at a chain restaurant near the station, where we had decided to stop for a bite after school. It was evening, bordering on night. It was raining outside. Still.

I'd spent the whole time after school helping Tetra, who was still worried about her presentation at the Narabikura Library conference. She had finally picked a topic, at least—algorithms. She was working hard to pull things together into a passable lecture, which was fine, but the amount... She had already written enough to fill an entire notebook, and was still at it.

"It's just too much," I said between mouthfuls of spaghetti. "No way you can cover all of this."

Tetra poked at her omelette. "But there's nothing I can cut! It all builds!"

"Even so, I'm pretty sure you'll have limited time."

Tetra looked at Lisa, who was sitting beside her. "Think I can get it extended?"

"No," Lisa said. She took a sip of her iced tea. As one of the coordinators, she would know.

"But if I don't tell it right, nobody will understand!"

"And nobody will understand if you cut your talk short because you've run out of time, so..."

"So I guess I'll just have to figure something out, yeah."

"You'd better," Lisa muttered. Tetra and I turned to look at her, expressionless and chewing on a straw. I wondered what had caused her to make such a cold comment.

"I... I'll do my best," Tetra said. "I just don't want anyone to walk away saying they didn't understand what they heard."

"Preening," Lisa said.

Tetra looked uncharacteristically miffed. "No, that's not it at all. I just want to get everything ready ahead of time."

Lisa glanced at me. "Using his time."

"Yes, yes I am, and I feel terrible about it." Tetra lowered her voice. "I know he needs to study for college entrance exams. I just want everything to be perfect..."

I decided it was time to step in between them. "Let's talk about something a bit more practical. You won't have much time to give your presentation, and we don't have long to get ready, so I think you need to limit yourself to maybe two sort algorithms."

"I guess," Tetra reluctantly agreed. "Bubble sort, and one other good one..."

"Quicksort," Lisa said.

"Perfect! Mr. Muraki just gave me a card about that one!"

Tetra started digging in her bag, but I held up a hand. "Sorry, but I'm exhausted. If I see that card it'll keep me up all night. Let's save it for tomorrow, after school. Day after tomorrow, maybe."

"Sorry again for taking up all your time," Tetra said.

Lisa said nothing, just sat there playing with her hair.

10.2 AT SCHOOL

10.2.1 Lunchtime

Two days later during lunch, I was talking with Miruka in our classroom. I told her about how hard Tetra was working to get ready for her presentation.

"Hmph," Miruka said, nibbling on a Kit Kat. "What about Lisa?"

"Lisa?" I grimaced. "Kinda tetchy, to be honest. She did say she'd help Tetra with her slides once she's written everything up, though."

"Sounds like her."

"Yeah. She's not your typical first-year student, is she?"

"In so many ways. By the way, are you going to the conference?"

"Of course. I'm taking Yuri too. You'll be out of the country, right?"

"I will."

"What are you going for this time?"

"A number theory conference. Should be fun. I'll be back in a week, the day after the Narabikura conference."

10.2.2 The Quicksort Algorithm

When my classes were done, I found Lisa and Tetra already in the library.

"We're talking about quicksort!" Tetra said. "I don't get it all quite yet, but Lisa and I have made some progress!"

"Well you're an eager pair," I said.

Tetra opened up her notebook. "We'll start with input and output."

The quicksort algorithm (input and output)

Input
- Array $A = \langle A[1], A[2], A[3], \ldots, A[n] \rangle$[1]
- Sort range, L to R

Output

An array with elements $A[L]$ to $A[R]$ sorted

"The quicksort algorithm takes an array A, divides it up into boxes, and sorts the part from L to R. Anything outside of that range is left alone. If you want to sort the entire array, just set L to 1 and R to n."

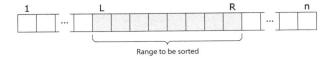

"Here's the procedure."

---

[1]The following analysis assumes all array elements have different values.

The quicksort algorithm (procedure)

```
R1:   procedure QUICKSORT(A, L, R)
R2:     if L < R then
R3:       p ← L
R4:       k ← L + 1
R5:       while k ⩽ R do
R6:         if A[k] < A[L] then
R7:           A[p + 1] ↔ A[k]
R8:           p ← p + 1
R9:         end-if
R10:        k ← k + 1
R11:      end-while
R12:      A[L] ↔ A[p]
R13:      A ← QUICKSORT(A, L, p − 1)
R14:      A ← QUICKSORT(A, p + 1, R)
R15:    end-if
R16:    return A
R17:  end-procedure
```

"The L and the R stand for 'left' and 'right'?" I asked.

"R!" Tetra said. I rolled my eyes. "I spent all day yesterday studying this algorithm. Lisa gave me lots of help, too. I wanted to get everything organized so that I don't waste your time." Tetra held up a bundle of papers, apparently notes she had taken. "Let me start by showing you a diagram of how the algorithm works."

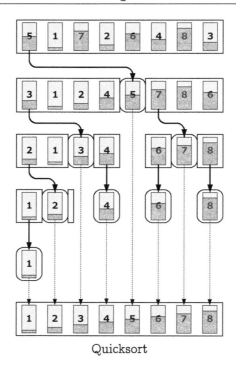

Quicksort

"This figure shows the QUICKSORT algorithm when the input is $A = \langle 5, 1, 7, 2, 6, 4, 8, 3 \rangle$, $L = 1$, $R = 8$. The first thing we do is select an element to be what's called the 'pivot.' That's going to serve as our basis for separating large and small values. QUICKSORT uses the leftmost value as the pivot, so in the figure we start out with a pivot value of 5. We move values smaller than the pivot to the left, and equal or larger values to the right. The pivot becomes like a separator between them. This is called partitioning."

"Huh," I said. "So in this case the partitions are $\langle 3, 1, 2, 4 \rangle$ and $\langle 7, 8, 6 \rangle$, with the pivot 5 in between?"

"Right! So now that we have two partitioned arrays, we perform a quicksort again on each of them. We sort the subarrays, in other words. So quicksort is basically a repetition of two steps."

- Partitioning with a pivot ($R3$ through $R12$)
- Sorting of the subarrays ($R13$ and $R14$)

"Now let me explain each in turn!"

10.2.3 Array Partitioning with a Pivot: Two Wings

"Okay, first let's talk about partitioning with a pivot," Tetra said. "We take the value in the leftmost box A[L] as the pivot, shown as an equals sign in the figure, meaning that value is equal to the pivot. We won't know if the other elements are larger or smaller than the pivot until we check them, so those are shown as a question mark. The variables p and k are initialized at lines $R3$ and $R4$, so things look like this."

Setting the value of A[L] as the pivot

"In the if statement on line $R6$, we compare an element with the pivot, and we repeat this processing in the while loop from line $R5$ to line $R11$."

$\vdots$

$R5$:   while $k \leqslant R$ do
$R6$:   if $A[k] < A[L]$ then
$R7$: $A[p+1] \leftrightarrow A[k]$
$R8$: $p \leftarrow p+1$
$R9$:   end-if
$R10$: $k \leftarrow k+1$
$R11$:   end-while

$\vdots$

If $A[2] < pivot$

"If $A[2]$ is less than the pivot value, we execute lines $R7$ through $R10$. Then once we return to line $R5$ from line $R11$, things look like this."

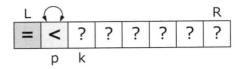

"The value in A[2] is 1 and the pivot value is 5, so the algorithm sees that A[2] is less than the pivot. That's shown in the figure as a < sign. The elements are exchanged in the A[p+1] ↔ A[k] swap at line *R7*. In this case it seems kind of silly to swap A[2] with itself, but that's just because p + 1 happens to equal k here. Everywhere else we'll be making more meaningful swaps for partitioning."

If A[2] ⩾ *pivot*

"When A[2] is at least the pivot value, p doesn't change and k increases."

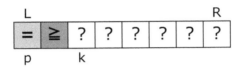

"So long as the k ⩽ R condition at *R5* holds, we just repeat the same thing, sorting out each element according to whether it's smaller than the pivot value. Here's what things look like partway through."

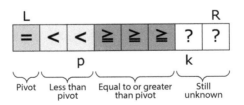

"This gives us a pretty clear picture of the current state of the partitioning."

- Elements less than the pivot value are in locations L+1 through p.

- Elements of at least the pivot value are in locations $p+1$ through $k-1$.

"The variable p marks the boundary between elements smaller and larger than the pivot, and k marks the front line of our partitioning task.

"So anyway, when k gets so large that $k \leqslant R$ no longer holds, the while loop ends."

$$
\begin{aligned}
&\vdots \\
R3:\quad &p \leftarrow L \\
R4:\quad &k \leftarrow L + 1 \\
R5:\quad &\text{while } k \leqslant R \text{ do} \\
&\vdots \\
R11:\quad &\text{end-while} \\
R12:\quad &A[L] \leftrightarrow A[p] \\
&\vdots
\end{aligned}
$$

"When the loop ends and we arrive at $R12$, we generally get something like this."

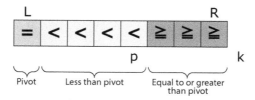

"From this state, we swap $A[L]$ and $A[p]$ at line $R12$."

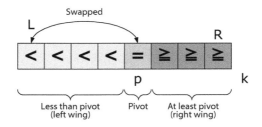

Two "wings"

"I'm going to call the elements to the left of the pivot the 'left wing,' and those to the right the 'right wing.' So we've split the array into left and right wings!"

Left wing = Set of elements less than the pivot value

Right wing = Set of elements with at least the pivot value

"However—and this is something that Lisa pointed out, thanks Lisa!—the pivot could end up at the far left or right. In that case one of the wings will disappear."

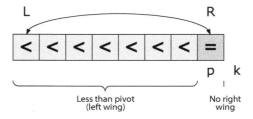

The right wing disappears when the pivot is at the far right.

"And I think that's about it for partitioning with a pivot!"

10.2.4 Sorting subarrays—Recursion

"So now let's talk about sorting the subarrays we've created," Tetra continued, "these left and right wings. If we sort the one on the left, then the one on the right, the whole array will be sorted!"

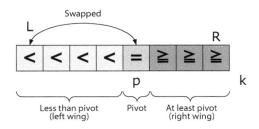

Sorting the two wings

"The left wing covers the range L to $p-1$, and the right wing covers $p+1$ to R. We sort these ranges at lines $R13$ and $R14$."

$$\vdots$$
$R13:$ $\quad$ A $\leftarrow$ QUICKSORT$(A, L, p-1)$
$R14:$ $\quad$ A $\leftarrow$ QUICKSORT$(A, p+1, R)$
$$\vdots$$

"As you can see, we're using calls to QUICKSORT in its own definition, which is called recursion. Thanks for pointing that out too, Lisa! It's kind of like a recurrence relation in math."

10.2.5 Analysis of Execution Steps

"Huh, this is pretty cool," I said. "Two sorts in one, the left and right wings to make up the whole."

"Divide and conquer," Lisa said.

"We didn't only look at how the quicksort algorithm works," Tetra said. "We also made a lot of progress on analyzing its number of execution steps."

Tetra turned and gave Lisa a conspiratorial smile. Lisa replied with a slight nod.

"As usual, we start by counting the number of times each line executes."

	Executions ($L \geqslant R$)	Executions ($L < R$)	Quicksort
$R1$:	1	1	procedure QUICKSORT(A, L, R)
$R2$:	1	1	if $L < R$ then
$R3$:	0	1	$p \leftarrow L$
$R4$:	0	1	$k \leftarrow L + 1$
$R5$:	0	$R - L + 1$	while $k \leqslant R$ do
$R6$:	0	$R - L$	if $A[k] < A[L]$ then
$R7$:	0	W	$A[p+1] \leftrightarrow A[k]$
$R8$:	0	W	$p \leftarrow p + 1$
$R9$:	0	W	end-if
$R10$:	0	$R - L$	$k \leftarrow k + 1$
$R11$:	0	$R - L$	end-while
$R12$:	0	1	$A[L] \leftrightarrow A[p]$
$R13$:	0	T_{left}	$A \leftarrow$ QUICKSORT$(A, L, p - 1)$
$R14$:	0	T_{right}	$A \leftarrow$ QUICKSORT$(A, p + 1, R)$
$R15$:	0	1	end-if
$R16$:	1	1	return A
$R17$:	1	1	end-procedure

Analysis of the QUICKSORT procedure

"There are some variables in the 'Executions' columns."

· R and L are given inputs.

· W is the number of times swaps occur at $R7$, but we don't know what that will be, since it depends on the input array.

· T_{left} is the number of steps we need to sort the left wing.

· T_{right} is the number of steps we need to sort the right wing.

"T_{left} and T_{right} will change according to the input, so I'm not sure how we can deal with that. I can explain how we got what we did figure out, though! First, let's use this to describe the number of QUICKSORT execution steps."

$$T_Q(R - L + 1)$$

"Then we get this, assuming $L < R$."

$$\begin{aligned}T_Q(R-L+1) &= R1+R2+R3+R4+R5+R6+R7+R8+R9\\&\quad+R10+R11+R12+R13+R14+R15+R16+R17\\&= 1+1+1+1+(R-L+1)+(R-L)+W+W+W\\&\quad+(R-L)+(R-L)+1+T_{\text{left}}+T_{\text{right}}+1+1+1\\&= 9+4R-4L+3W+T_{\text{left}}+T_{\text{right}}\end{aligned}$$

"If instead $L \geqslant R$, specifically if the size is 0 because $R-L+1=0$ or when the size is 1 because $R-L+1=1$, we get this."

$$T_Q(0) = R1+R2+R16+R17 = 1+1+1+1 = 4$$
$$T_Q(1) = R1+R2+R16+R17 = 1+1+1+1 = 4$$

"And there you have it! That's as far as Lisa and I got."

Quicksort analysis by Tetra and Lisa

Given $L < R$, the QUICKSORT algorithm requires

$$T_Q(R-L+1) = 9+4R-4L+3W+T_{\text{left}}+T_{\text{right}}$$

steps to sort an array from L to R, where

- W is the number of element exchanges at line $R7$,
- T_{left} is the number of steps in the left wing, and
- T_{right} is the number of steps in the right wing.

10.2.6 By Cases

"So there's what we have so far," Tetra said.

$$T_Q(R-L+1) = 9+4R-4L+3W+T_{\text{left}}+T_{\text{right}}$$

Lisa nodded in agreement. I noticed something unusual—she had been paying full attention to what Tetra was saying, without even opening her computer. I could empathize, though; Tetra's description had really interested me, too. I was already wondering how we might better evaluate this equation they had come up with.

"I wonder if we can't get rid of some of these letters," I said.

"How do you mean?"

"Well, when you sort from L to R, the sort involves $R - L + 1$ elements, right? I assume that's why you're writing the number of steps as $T_Q(R - L + 1)$."

"That's right. Because the number of elements will be one more than the right side minus the left side."

"So that means we can let n be the number of elements, and write $T_Q(R - L + 1)$ as $T_Q(n)$ instead. And if we're going to do that, it seems like we can write T_{left} and T_{right} in a $T_Q(\cdots)$ form too."

"We can! Since the left-wing sort is QUICKSORT$(A, L, p - 1)$ we can write it as

$$\begin{array}{c}\text{Number of}\\ \text{left-wing elements}\end{array} = \text{right} - \text{left} + 1 = (p - 1) - L + 1 = p - L,$$

and the right-wing sort is QUICKSORT$(A, p + 1, R)$, so we get

$$\begin{array}{c}\text{Number of}\\ \text{right-wing elements}\end{array} = \text{right} - \text{left} + 1 = R - (p + 1) + 1 = R - p,$$

and in the end we get this."

$$\begin{cases} T_{\text{left}} & = T_Q(p - L) \\ T_{\text{right}} & = T_Q(R - p) \end{cases}$$

Something about this bothered me. "Hmm. You sure about that?"

"Sure! Then we can reduce the number of variables like this."

$$T_Q(R - L + 1) = 9 + 4R - 4L + 3W + T_{\text{left}} + T_{\text{right}}$$
$$T_Q(n) = 9 + 4R - 4L + 3W + T_Q(p - L) + T_Q(R - p)$$

"Same thing," Lisa said.

Tetra did a double-take. "Huh? We had $R, L, W, T_{\text{left}}, T_{\text{right}}$, and now we have n, R, L, W, p, so..."

"If you simplify the $9 + 4R - 4L$ as $R - L + 1$ you can represent it as n," I said.

$$\begin{aligned} T_Q(n) &= 9 + 4R - 4L + 3W + T_Q(p - L) + T_Q(R - p) \\ &= 4(\underline{R - L + 1}) + 5 + 3W + T_Q(p - L) + T_Q(R - p) \\ &= 4\underline{n} + 5 + 3W + T_Q(p - L) + T_Q(R - p) \end{aligned}$$

"What about the W?" Tetra said.

"Good question. Well, we're after the number of execution steps, so how about if we evaluate W a bit? W is the number of swaps, right? The if statement in $R6$ comes $R - L$ times, so W can't be any more than that."

	Executions $(L \geqslant R)$	Executions $(L < R)$	Quicksort
⋮			
$R6$:	0	$R - L$	if $A[k] < A[L]$ then
$R7$:	0	W	$A[p+1] \leftrightarrow A[k]$
$R8$:	0	W	$p \leftarrow p + 1$
$R9$:	0	W	end-if
⋮			

"Oh, I see!" Tetra said.

"So let's estimate a largish W, as $W = R - L = n - 1$."

$$\begin{aligned} T_Q(n) &= 4n + 5 + 3\underline{W} + T_Q(p - L) + T_Q(R - p) \\ &= 4n + 5 + 3\underline{(R - L)} + T_Q(p - L) + T_Q(R - p) \\ &= 4n + 5 + 3\underline{(n - 1)} + T_Q(p - L) + T_Q(R - p) \\ &= 7n + 2 + T_Q(p - L) + T_Q(R - p) \end{aligned}$$

"Wow, much simpler!"

$$T_Q(n) = 7n + 2 + T_Q(p - L) + T_Q(R - p)$$

"We're erring on the side of caution here, so strictly speaking $T_Q(n)$ is showing the number of execution steps for a somewhat inefficient algorithm. But even so, something's still bothering me... p is where the pivot ends up being, right? So p depends on the input array A. Something about that..."

"Meaning p can take lots of different values?"

"Right. I know that p will have to take a value in the range $L \leqslant p \leqslant R$, so the range for $p - L$ is

$$0 \leqslant p - L \leqslant R - L = n - 1,$$

and the range for $R - p$ is

$$0 \leqslant R - p \leqslant R - L = n - 1,$$

but still, something's off."

"Because you can't actually know which is the case?"

"Right. Considering things by cases is such a pain."

"I like it," Lisa said.

"You like it?" Tetra asked, to which Lisa nodded.

"Well, we have to do it anyway, like it or not," I said. "Even if we've settled on an n, if we don't have a fixed p then $T_Q(n)$ won't settle on a specific value, and that won't do. Since $0 \leqslant p - L \leqslant n - 1$, there are n cases for $p - L$. We end up with a recurrence relation that depends on pivot location."

"I guess we do," Tetra said.

"I think it's time to start laying some breadcrumbs, in case we lose our way. Let me write a recurrence relation."

A breadcrumb for QUICKSORT

$$\begin{cases} T_Q(0) &= 4 \\ T_Q(1) &= 4 \\ T_Q(n) &= 7n + 2 + T_Q(p - L) + T_Q(R - p) \quad (n = 2, 3, 4, \ldots) \end{cases}$$

(but we still have this variable p...)

10.2.7 Maximum Execution Steps

"You want to get rid of the division into cases according to p, right?" Tetra said.

I nodded. "Right."

"So one thing I thought of, maybe we can consider the maximum case, like before when we said W could be at most $R - L$. We could say that every time we partition, the pivot is at the leftmost side. Then we would have $p = L$!"

"Interesting. So let's write that worst-case scenario for $T_Q(n)$ as $T'_Q(n)$."

$$T'_Q(n) = \text{Maximum number of QUICKSORT steps}$$

Tetra hurriedly developed this into an equation.

$$\begin{aligned}
T'_Q(n) &= 7n + 2 + T'_Q(p - L) + T'_Q(R - p) \\
&= 7n + 2 + T'_Q(L - L) + T'_Q(R - L) && \text{let } p = L \\
&= 7n + 2 + T'_Q(0) + T'_Q(n - 1) && \text{calculate} \\
&= 7n + 2 + 4 + T'_Q(n - 1) && \text{use } T'_Q(0) = 4 \\
&= T'_Q(n - 1) + 7n + 6 && \text{rearrange}
\end{aligned}$$

"So here's the recurrence relation."

$$\begin{cases} T'_Q(0) &= 4 \\ T'_Q(1) &= 4 \\ T'_Q(n) &= T'_Q(n-1) + 7n + 6 \end{cases} \quad (n = 2, 3, 4, \ldots)$$

"Well done, Tetra!" I said. "We should be able to solve this easily. I bet we'll find a pattern if we make things smaller, going $n, n-1, n-2, \ldots$"

$$\begin{aligned}
T'_Q(n) &= \underline{T'_Q(n-1)} + 7n + 6 \\
&= \underline{T'_Q(n-2) + 7(n-1) + 6} + 7n + 6 \\
&= \underline{T'_Q(n-2)} + \bigl(7(n-1) + 6\bigr) + (7n + 6) \\
&= \underline{T'_Q(n-3) + 7(n-2) + 6} + \bigl(7(n-1) + 6\bigr) + (7n + 6) \\
&= T'_Q(n-3) + \bigl(7(n-2) + 6\bigr) + \bigl(7(n-1) + 6\bigr) + (7n + 6)
\end{aligned}$$

"The pattern is easier to see if we write n as n − 0."

$$T'_Q(n) = T'_Q(n-3) + \big(7\underline{(n-2)} + 6\big) + \big(7\underline{(n-1)} + 6\big) + \big(7\underline{(n-0)} + 6\big)$$

"Let's write this as a sum."

$$= T'_Q(n-3) + \sum_{j=n-2}^{n-0} \big(7j + 6\big)$$

"Now we can make this smaller, $n-4, n-5, \ldots, n-k$"

$$T'_Q(n) = T'_Q(n-k) + \sum_{j=n-k+1}^{n} \big(7j + 6\big)$$

"We end up at $n - (n-1)$, in other words 1."

$$T'_Q(n) = T'_Q(1) + \sum_{j=2}^{n} \big(7j + 6\big)$$

$$= 4 + \sum_{j=2}^{n} \big(7j + 6\big) \qquad \text{use } T'_Q(1) = 4$$

"I'm going to organize things so that the sum starts from $j = 1$."

$$T'_Q(n) = 4 - (7 \cdot 1 + 6) + \sum_{j=1}^{n} \big(7j + 6\big)$$

$$= -9 + \sum_{j=1}^{n} 7j + \sum_{j=1}^{n} 6$$

$$= -9 + 7\sum_{j=1}^{n} j + 6n$$

$$= 6n - 9 + 7\sum_{j=1}^{n} j$$

$$= 6n - 9 + \frac{7n(n+1)}{2} \qquad (\text{use } 1 + 2 + 3 + \cdots + n = \frac{n(n+1)}{2})$$

"Reorganizing to put things in terms of n, we get this."

$$T'_Q(n) = \frac{7}{2}n^2 + \frac{19}{2}n - 9$$

"In other words, using big-O notation we get this."

$$T'_Q(n) = O(n^2)$$

"Hang on," Tetra said. "So it's at most order n^2? Just like a bubble sort?"

"When we've added on the condition of leftmost pivot, yeah."

"Wait, wait, wait. We're selecting the leftmost value as the pivot every time, right? Which means that every time the pivot is the smallest value, which means... Doesn't that mean the array is *already sorted*?"

"Hey, you're right."

"So the order of the maximum number of quicksort execution steps is $O(n^2)$ when it's given an already sorted array."

$$T'_Q(n) = O(n^2)$$

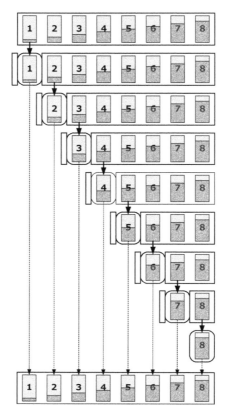

Quicksort on an already sorted array.

"But now it looks like we're stuck," Tetra said. "I thought we could get rid of the different cases if we fixed the position of the pivot, but..." She gave her cheek a twist. "Alright, so there are n possibilities for the pivot's location, 1 through n. And I guess there's no way we could calculate that many values all at once."

"I guess not. Looks like we're going to have to generalize to account for the n possibilities."

"When we have to handle lots of values..." Tetra chanted.

"What's that?"

"Something Miruka said the other day. 'When we have to handle lots of values, it's natural to want to summarize them.' I was just wondering how we can summarize these."

"She did say that, didn't she. Something about...averages, wasn't it?"

"Oh!"

"Of course!"

"An average!"

Lisa said nothing, but her eyes spoke volumes.

"Okay, new goal," I said. "We don't want the number of steps, we want the *average* number of steps. Let's find it!"

10.2.8 Average Execution Steps

I opened to a new page in my notebook. "Okay, let's figure out the average number of execution steps in the quicksort algorithm. The average for all possible inputs of size n. Let's call that $\bar{T}_Q(n)$."

$$\bar{T}_Q(n) = \text{Average number of execution steps}$$

"We'll start over from our last breadcrumb, and create a new recurrence relation for the average number of steps."

"Okay, but there's infinitely many possible inputs for a size of n. There are no limits on the range for the elements, so it could be something like $\langle 5, 1, 7, 2, 6, 4, 8, 3 \rangle$, or it could be $\langle 500, 100, 700, 200, 600, 400, 800, 300 \rangle$, or anything! How do you average infinity?"

"No worries, we don't have to think about the values of the elements per se. We just need to consider size relations between elements. For simplicity, let's say all the elements are different, and that we're sorting n elements $1, 2, 3, \ldots, n$."

"Okay, but I still don't see how we're pulling an average out of that."

"It's not hard. If the input is an array of size n, there are n! ways to arrange the numbers in it. Since we're considering all of those arrangements, the pivot will always be located at some position 1 through n. So let's consider p as moving through the entire range, $p = L, L+1, L+2, \ldots, R-1, R$. There are $R-L+1$, in other words n, locations for p. To get an average, we add them all up and divide by n. To do that, we can create a recurrence relation."

> **A recurrence relation for QUICKSORT ($\bar{T}_Q(n)$ is the average number of execution steps)**
>
> $$\begin{cases} \bar{T}_Q(0) &= 4 \\ \bar{T}_Q(1) &= 4 \\ \bar{T}_Q(n) &= \dfrac{1}{n}\sum_{p=L}^{R}\left(7n+2+\bar{T}_Q(p-L)+\bar{T}_Q(R-p)\right) \end{cases}$$
>
> $(n = 2, 3, 4, \ldots)$

"Let's let $j = p - L + 1$, to make everything easier to read. We'll also use $p-L = j-1$ and $R-p = R-(j+L-1) = (R-L+1)-j = n-j$."

$$\bar{T}_Q(n) = \frac{1}{n}\sum_{j=1}^{n}\left(7n+2+\bar{T}_Q(j-1)+\bar{T}_Q(n-j)\right)$$

"First off, let's get rid of the sigma and get things into the form of a recurrence relation for $\bar{T}_Q(n)$. We can start by moving $7n + 2$ to the outside, since it doesn't rely on j. Since the sigma is adding everything up n times from 1 to n, when we move it to the outside we need to multiply it by n. It was already being multiplied by $\frac{1}{n}$, so the n's just disappear."

$$\begin{aligned}
\bar{T}_Q(n) &= \frac{1}{n}\sum_{j=1}^{n}\left(\underline{7n+2}+\bar{T}_Q(j-1)+\bar{T}_Q(n-j)\right) \\
&= \frac{1}{\not{n}}\cdot\not{n}\cdot\underline{(7n+2)}+\frac{1}{n}\sum_{j=1}^{n}\left(\bar{T}_Q(j-1)+\bar{T}_Q(n-j)\right) \\
&= 7n+2+\frac{1}{n}\sum_{j=1}^{n}\left(\bar{T}_Q(j-1)+\bar{T}_Q(n-j)\right) \\
&= 7n+2+\frac{1}{n}\sum_{j=1}^{n}\bar{T}_Q(j-1)+\frac{1}{n}\sum_{j=1}^{n}\bar{T}_Q(n-j)
\end{aligned}$$

"The results of the two sums here will be equal. We can tell that's true, because one is just the other in reverse. Like this."

$$\begin{cases} \sum_{j=1}^{n} \bar{T}_Q(j-1) = \bar{T}_Q(0) + \bar{T}_Q(1) + \bar{T}_Q(2) + \cdots + \bar{T}_Q(n-1) \\ \sum_{j=1}^{n} \bar{T}_Q(n-j) = \bar{T}_Q(n-1) + \cdots + \bar{T}_Q(2) + \bar{T}_Q(1) + \bar{T}_Q(0) \end{cases}$$

"This means we can combine the two sums into one."

$$\bar{T}_Q(n) = 7n + 2 + \frac{1}{n}\sum_{j=1}^{n}\bar{T}_Q(j-1) + \frac{1}{n}\sum_{j=1}^{n}\bar{T}_Q(n-j)$$

$$= 7n + 2 + \frac{2}{n}\sum_{j=1}^{n}\bar{T}_Q(j-1) \qquad \text{combined, coefficient doubled}$$

"I'd like to get rid of this $\frac{2}{n}$, so to do that I'm going to multiply both sides by n to remove the denominator."

$$n \cdot \bar{T}_Q(n) = 7n^2 + 2n + 2\sum_{j=1}^{n}\bar{T}_Q(j-1)$$

"Much better. But still, we want this to be in the form $\bar{T}_Q(n)$, so we need to do something about the sigma on the right. How to do that, though... Ah, of course. The usual way of solving recurrence relations, a difference. We just need to calculate $(n+1) \cdot \bar{T}_Q(n+1) -$

$n \cdot \bar{T}_Q(n)$, that will do it."

$$(n+1) \cdot \bar{T}_Q(n+1) - n \cdot \bar{T}_Q(n)$$
$$= \left(7(n+1)^2 + 2(n+1) + 2\sum_{j=1}^{n+1} \bar{T}_Q(j-1)\right)$$
$$- \left(7n^2 + 2n + 2\sum_{j=1}^{n} \bar{T}_Q(j-1)\right)$$
$$= \left(\cancel{7n^2} + 14n + 7 + \cancel{2n} + 2 + 2\sum_{j=1}^{n+1} \bar{T}_Q(j-1)\right)$$
$$- \left(\cancel{7n^2} + \cancel{2n} + 2\sum_{j=1}^{n} \bar{T}_Q(j-1)\right)$$
$$= 14n + 9 + 2\left(\sum_{j=1}^{n+1} \bar{T}_Q(j-1) - \sum_{j=1}^{n} \bar{T}_Q(j-1)\right)$$
$$= 14n + 9 + 2 \cdot \bar{T}_Q((n+1)-1)$$
$$= 14n + 9 + 2 \cdot \bar{T}_Q(n)$$

Tetra raised a hand. "How did you get rid of the sigmas here?"
"Like this."

$$\sum_{j=1}^{n+1} \bar{T}_Q(j-1) = \underbrace{\bar{T}_Q(0) + \bar{T}_Q(1) + \cdots + \bar{T}_Q(n-1)}_{\sum_{j=1}^{n} \bar{T}_Q(j-1)} + \bar{T}_Q(n)$$

"If we subtract the sum for $j = 1, \ldots, n$ from the sum for $j = 1, \ldots, n, n+1$, only the $j = n+1$ term will remain."

"Much better, thanks!"

"Okay, now that we have terms including $\bar{T}_Q(n)$ on both sides, lets move $n \cdot \bar{T}_Q(n)$ to the right."

$$(n+1) \cdot \bar{T}_Q(n+1) - n \cdot \bar{T}_Q(n) = 14n + 9 + 2 \cdot \bar{T}_Q(n)$$
$$(n+1) \cdot \bar{T}_Q(n+1) = n \cdot \bar{T}_Q(n) + 14n + 9 + 2 \cdot \bar{T}_Q(n)$$
$$(n+1) \cdot \bar{T}_Q(n+1) = (n+2) \cdot \bar{T}_Q(n) + 14n + 9$$

"From what we've done so far, we found that this recurrence relation holds for $n = 2, 3, \ldots$. See how the sigmas have disappeared?"

$$(n+1) \cdot \bar{T}_Q(n+1) = (n+2) \cdot \bar{T}_Q(n) + 14n + 9$$

"We should also see what happens when $n = 2$."

$$\bar{T}_Q(n) = \frac{1}{n} \sum_{j=1}^{n} \left(7n + 2 + \bar{T}_Q(j-1) + \bar{T}_Q(n-j) \right)$$

$$\bar{T}_Q(2) = \frac{1}{2} \sum_{j=1}^{2} \left(7 \cdot 2 + 2 + \bar{T}_Q(j-1) + \bar{T}_Q(2-j) \right)$$

$$= \frac{1}{2} \left(\left(7 \cdot 2 + 2 + \bar{T}_Q(0) + \bar{T}_Q(1) \right) + \left(7 \cdot 2 + 2 + \bar{T}_Q(1) + \bar{T}_Q(0) \right) \right)$$

$$= 14 + 2 + \bar{T}_Q(0) + \bar{T}_Q(1)$$

$$= 14 + 2 + 4 + 4$$

$$= 24$$

"From here, we need to move on to asymptotic analysis, so—"

Tetra held up both hands. "Whoa! I know equations energize you, but they exhaust me."

"Yeah, you have a point there."

"So this $\bar{T}_Q(n)$, we still don't know the average number of execution steps for quicksort?"

"Well, quicksort is a comparative algorithm, so it will have to be of order $n \log n$ at the very least. I wonder if there's some way we can hold it down to $n \log n$ from above too?"

"The library is *closed!*" came Ms. Mizutani's call.

Problem 10-1 (Average quicksort execution steps)

The average number of QUICKSORT execution steps $\bar{T}_Q(n)$ fulfills the recurrence relation

$$\begin{cases} \bar{T}_Q(0) &= 4 \\ \bar{T}_Q(1) &= 4 \\ \bar{T}_Q(2) &= 24 \\ (n+1) \cdot \bar{T}_Q(n+1) &= (n+2) \cdot \bar{T}_Q(n) + 14n + 9 \end{cases}$$

for $n = 2, 3, 4, \ldots$. Given that, does the following equation hold?

$$\bar{T}_Q(n) = O(n \log n)$$

10.2.9 On the Way Home

Tetra, Lisa, and I headed off to the train station.

"So you're going to talk about bubble sort and quicksort?" I asked Tetra.

"Yeah. But I'm trying to set it up in a way that shows how fun it is to discover how they work yourself. Ironic, I guess, since you and Lisa helped me so much with their analyses."

I reflected on my own recent discoveries, not only how surprisingly stubborn Tetra could be, but also how the seemingly cold Lisa was so eager to help, silently supporting Tetra in preparing for the conference.

"Miruka never showed up today, did she," Tetra said.

I shrugged. "She must have been busy with something."

"And she won't be at the conference?"

"That's what she says. She'll be in the States."

"Oh, I see."

Lisa remained silent.

10.3 AT HOME

10.3.1 Changing Forms

That Saturday, Yuri came to "study," so I told her about our quicksort analysis.

Problem 10-1 (Average quicksort execution steps)

The average number of QUICKSORT execution steps $\bar{T}_Q(n)$ fulfills the recurrence relation

$$\begin{cases} \bar{T}_Q(0) & = 4 \\ \bar{T}_Q(1) & = 4 \\ \bar{T}_Q(2) & = 24 \\ (n+1) \cdot \bar{T}_Q(n+1) & = (n+2) \cdot \bar{T}_Q(n) + 14n + 9 \end{cases}$$

for $n = 2, 3, 4, \ldots$. Given that, does the following equation hold?

$$\bar{T}_Q(n) = O(n \log n)$$

"Above my pay grade," Yuri said.

"Well, it wouldn't be so bad if the function was in some simple form like this."

$$\begin{cases} f(1) & = <\text{something}> \\ f(n+1) & = f(n) + <\text{something}> \end{cases} \quad (n = 1, 2, 3, \ldots)$$

"Then we could just make the n in the f(n) smaller until we got to where we want to go. But in this recurrence relation we've got $\bar{T}_Q(n)$ mixed in with these other bits that have n's in them, so..."

Yuri rolled her eyes. "If the n bits are a problem, just get rid of them."

"Easier said than done. I mean— Oh, hold on."

"What?"

"Maybe it's not so bad after all. Look what happens if we divide both sides by $(n+1)(n+2)$."

$(n+1) \cdot \bar{T}_Q(n+1) = (n+2) \cdot \bar{T}_Q(n) + 14n + 9$ the recurrence relation

$$\frac{\bar{T}_Q(n+1)}{n+2} = \frac{\bar{T}_Q(n)}{n+1} + \frac{14n+9}{(n+1)(n+2)} \quad \text{divide by } (n+1)(n+2)$$

"Wow, that's...a total mess. How am I supposed to read this?"

"With creative definitions. Something like this."

$$F(n) = \frac{\bar{T}_Q(n)}{n+1}$$

"Now we can represent things using $F(n)$."

$$\frac{\bar{T}_Q(n+1)}{n+2} = \frac{\bar{T}_Q(n)}{n+1} + \frac{14n+9}{(n+1)(n+2)} \quad \text{from above}$$

$$F(n+1) = F(n) + \frac{14n+9}{(n+1)(n+2)} \quad \text{represent as } F(n)$$

"That's an improvement...I guess? But you've still got that freaky fraction."

"I think we can break the freaky fraction down into a simple sum. Let's see... Say we've used a and b to decompose things into a sum, like this."

$$\frac{14n+9}{(n+1)(n+2)} = \frac{a}{n+1} + \frac{b}{n+2}$$

"Calculating this, we get...

$$\frac{a}{n+1} + \frac{b}{n+2} = \frac{(a+b)n + (2a+b)}{(n+1)(n+2)}$$

"... which means this holds."

$$\frac{\boxed{14}n + \boxed{9}}{(n+1)(n+2)} = \frac{(\boxed{a+b})n + (\boxed{2a+b})}{(n+1)(n+2)}$$

"Now we just need to solve this system of equations."

$$\begin{cases} a+b = 14 \\ 2a+b = 9 \end{cases}$$

"The solution is $(a, b) = (-5, 19)$, so that gives us this."

$$\frac{14n+9}{(n+1)(n+2)} = \frac{-5}{n+1} + \frac{19}{n+2}$$

"Now I want to get back to the F(n + 1) equation."

$$F(n+1) = F(n) + \frac{14n+9}{(n+1)(n+2)}$$

"And now we can change your freaky fraction into a sum."

$$F(n+1) = F(n) + \frac{-5}{n+1} + \frac{19}{n+2}$$

"Let's use the recurrence relation to put F(n) into a simpler form. That will make any patterns easier to see."

$$\begin{aligned}
F(n) &= \underwave{F(n-1)} + \frac{-5}{n-0} + \frac{19}{n+1} \\
&= F(n-2) + \frac{-5}{n-1} + \underwave{\frac{19}{n-0} + \frac{-5}{n-0}} + \frac{19}{n+1} \\
&= F(n-2) + \frac{-5}{n-1} + \left(\frac{19}{n-0} + \frac{-5}{n-0}\right) + \frac{19}{n+1} \\
&= \underwave{F(n-2)} + \frac{-5}{n-1} + \frac{14}{n-0} + \frac{19}{n+1} \\
&= F(n-3) + \frac{-5}{n-2} + \underwave{\frac{19}{n-1} + \frac{-5}{n-1}} + \frac{14}{n-0} + \frac{19}{n+1} \\
&= F(n-3) + \frac{-5}{n-2} + \left(\frac{19}{n-1} + \frac{-5}{n-1}\right) + \frac{14}{n-0} + \frac{19}{n+1} \\
&= \underwave{F(n-3)} + \frac{-5}{n-2} + \frac{14}{n-1} + \frac{14}{n-0} + \frac{19}{n+1} \\
&= F(n-4) + \frac{-5}{n-3} + \underwave{\frac{19}{n-2} + \frac{-5}{n-2}} + \frac{14}{n-1} + \frac{14}{n-0} + \frac{19}{n+1} \\
&= F(n-4) + \frac{-5}{n-3} + \left(\frac{19}{n-2} + \frac{-5}{n-2}\right) + \frac{14}{n-1} + \frac{14}{n-0} + \frac{19}{n+1} \\
&= F(n-4) + \frac{-5}{n-3} + \underbrace{\frac{14}{n-2} + \frac{14}{n-1} + \frac{14}{n-0}}_{\text{a pattern!}} + \frac{19}{n+1}
\end{aligned}$$

"Factoring out a 14 ..."

$$= F(n-4) + \frac{-5}{n-3} + 14\left(\frac{1}{n-2} + \frac{1}{n-1} + \frac{1}{n-0}\right) + \frac{19}{n+1}$$

"... and writing as a sum."

$$= F(n-4) + \frac{-5}{n-3} + 14 \underbrace{\sum_{j=n-2}^{n} \frac{1}{j}} + \frac{19}{n+1}$$

"Now I want to keep replacing until the $F(n-4)$ on the right becomes $F(2)$."

$$F(n) = F(n-(n-2)) + \frac{-5}{n-(n-2)+1} + 14 \sum_{j=n-(n-2)+2}^{n} \frac{1}{j} + \frac{19}{n+1}$$

$$= F(2) + \frac{-5}{3} + 14 \sum_{j=4}^{n} \frac{1}{j} + \frac{19}{n+1}$$

$$= F(2) + \frac{-5}{3} - 14 \left(\frac{1}{1} + \frac{1}{2} + \frac{1}{3} \right) + 14 \sum_{j=1}^{n} \frac{1}{j} + \frac{19}{n+1}$$

"Here we can use the fact that $F(2) = \frac{\bar{T}_Q(2)}{2+1} = \frac{24}{3} = 8$."

$$F(n) = \underbrace{8 + \frac{-5}{3} - 14 \left(\frac{1}{1} + \frac{1}{2} + \frac{1}{3} \right)}_{\text{constant part}} + 14 \sum_{j=1}^{n} \frac{1}{j} + \frac{19}{n+1}$$

$$= K + 14 \sum_{j=1}^{n} \frac{1}{j} + \frac{19}{n+1}$$

"Let's just define that constant part as K."

$$K = 8 + \frac{-5}{3} - 14 \left(\frac{1}{1} + \frac{1}{2} + \frac{1}{3} \right)$$

"Okay, back to $F(n)$."

$$F(n) = K + \frac{19}{n+1} + 14 \sum_{j=1}^{n} \frac{1}{j}$$

$$= K + \frac{19}{n+1} + 14 H_n$$

"Hold up," Yuri said. "What's with this H_n you're sneaking in to replace the sigma?"

"That's an old friend, a harmonic number. Here's how it's defined."

$$H_n = \frac{1}{1} + \frac{1}{2} + \frac{1}{3} + \cdots + \frac{1}{n} = \sum_{j=1}^{n} \frac{1}{j}$$

"Oh. Okay."

"Now all that's left is some mechanical calculations."

$$F(n) = \frac{\bar{T}_Q(n)}{n+1}$$

"So now we can represent $\bar{T}_Q(n)$ in terms of $F(n)$."

$$\begin{aligned}
\bar{T}_Q(n) &= (n+1) \cdot F(n) \\
&= (n+1) \cdot \left(K + \frac{19}{n+1} + 14 H_n\right) \\
&= K \cdot (n+1) + 19 + 14(n+1) H_n \\
&= \underbrace{14n \cdot H_n}_{\text{asymptotically large term}} + K \cdot n + 14 H_n + K + 19 \\
&= O(n \cdot H_n)
\end{aligned}$$

"And there we have it!"

$$\bar{T}_Q(n) = O(n \cdot H_n)$$

The average number of QUICKSORT execution steps $\bar{T}_Q(n)$ is

$$\bar{T}_Q(n) = O(n \cdot H_n),$$

where H_n is a harmonic number defined as

$$H_n = \frac{1}{1} + \frac{1}{2} + \frac{1}{3} + \cdots + \frac{1}{n}.$$

"Hey, you aren't done yet!" Yuri said. "You didn't want $\bar{T}_Q(n) = O(n \cdot H_n)$, you wanted $\bar{T}_Q(n) = O(n \log n)$, right?"

"Yeah, but harmonic numbers are asymptotically of the same order as a logarithmic function."

$$H_n = O(\log n) \quad \text{and} \quad n \cdot H_n = O(n \log n)$$

"So we can also say that $\bar{T}_Q(n) = O(n \log n)$."

Answer 10-1 (Average quicksort execution steps)

The order of the average number of QUICKSORT execution steps is

$$\bar{T}_Q(n) = O(n \log n).$$

10.3.2 H_n and $\log n$

Yuri still had a dissatisfied look. "Okay, I get that a harmonic number is a sum like $\frac{1}{1}, \frac{1}{2}, \frac{1}{3}, \cdots, \frac{1}{n}$, we've talked about that. But I still don't know what this $\log n$ is all about."

"The one in the $H_n = O(\log n)$?"

"Right. I've never seen that in school, so that's not fair."

"It's math, so it's fair, whether you've studied it or not. Anyway, the fact that H_n can be constrained to $\log n$ proves that $\sum \frac{1}{k}$ is constrained to $\int \frac{1}{x} dx$, if you'll excuse the integral."

"Even worse!"

"Maybe things will be clearer from a graph."

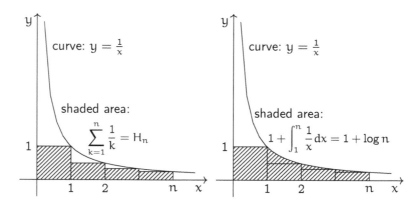

"That's a harmonic number on the left, a logarithmic function on the right," I said.

"And the shaded area on the right will always be larger than the shaded area on the left."

"Correct. We get the area on the right from an integral, and that's always going to be bigger, so the inequality holds."

"Integrals and areas? Still no idea."

"We'll study that someday."

"Ugh. By the way, what about preconditions?"

"Preconditions?"

"Yeah, you're the one that was going on about 'quantitative evaluation with clear preconditions' the other day. Aren't there any preconditions for finding the execution steps?"

"Not in this case, no. We're just looking for a simple average."

10.4 IN THE LIBRARY

10.4.1 Miruka

"There's a huge precondition," Miruka said.

On Monday we were back in the school library, where I was telling the usual gang about $\bar{T}_Q(n)$.

$$\bar{T}_Q(n) = O(n \log n)$$

"Seriously?" I said. "Like what?"

"You found the average number of execution steps. When you do that, you have to assume that all inputs are provided with equal probability. There's your huge precondition, that the probability distribution for the input is uniform."

"Huh."

"That's what allowed you to assert that the pivot position can be anywhere from 1 to n with equal probability, and that's why you were able to find the average by dividing the sum of execution steps by n."

"I guess that does make sense," I admitted.

"But...but..." Tetra stammered. "A uniform distribution? Doesn't that seem like a perfectly reasonable thing to assume?"

Miruka curled a lock of hair around a finger. "There's nothing wrong with making assumptions. You just need to note that they're there. If you're going to say that quicksort executes an order of $O(n \log n)$ steps on average, you also have to say you're assuming input with a uniform probability distribution."

"Duly noted."

"Say you run quicksort on an array that's already been sorted. You couldn't say you'll finish after $O(n \log n)$ steps, right?"

Tetra shook her head. "No, you couldn't. More like $O(n^2)$ steps, I'd think, if it was already sorted."

"So I guess it's better to have as few preconditions on an algorithm as possible," I said.

"Depends on how you define 'better.' I suppose it's better to be able to say 'for input with an arbitrary probability distribution' than 'for input with a uniform probability distribution,' in the sense that it allows for broader applications."

"That sounds like a bit of a stretch, though. How could you ever analyze an algorithm without worrying about the probability distribution of its input?"

"By using a randomized algorithm. That sometimes lets you bring in evaluations that don't rely on the probability distribution of its input. Like the RANDOM-WALK-3-SAT algorithm we talked about the other day. We were able to evaluate the probability of success, and to asymptotically analyze the number of execution steps."

Tetra grasped both sides of her head. "But Miruka, a random walk and sorting arrays... aren't those completely different things? Sorting is all about tidying up things that've been scattered about. How can you do that with a randomized algorithm?"

Miruka raised a finger and smiled. "With a randomized quicksort, of course."

10.4.2 Randomized Quicksort

"In a randomized quicksort, we randomly select the pivot," Miruka said. "The number of execution steps will vary according to the selected pivot, but the sort will complete all the same. So there isn't any particular problem with using random numbers to select a pivot."

"I guess not," Tetra said.

"Randomly selecting the pivot is all it takes to change quicksort to a randomized quicksort."

The randomized quicksort algorithm (procedure)

▶ *R1a*: procedure RANDOMIZED-QUICKSORT(A, L, R)
  R2:   if L < R then
▶ *R2a*: $r \leftarrow$ RANDOM(L, R)
▶ *R2b*: $A[L] \leftrightarrow A[r]$
  R3: $p \leftarrow L$
  R4: $k \leftarrow L + 1$
  R5:   while $k \leqslant R$ do
  R6:   if $A[k] < A[L]$ then
  R7: $A[p+1] \leftrightarrow A[k]$
  R8: $p \leftarrow p + 1$
  R9:   end-if
  R10: $k \leftarrow k + 1$
  R11:   end-while
  R12: $A[L] \leftrightarrow A[p]$
▶ *R13a*: $A \leftarrow$ RANDOMIZED-QUICKSORT$(A, L, p - 1)$
▶ *R14a*: $A \leftarrow$ RANDOMIZED-QUICKSORT$(A, p + 1, R)$
  R15:   end-if
  R16:   return A
  R17: end-procedure

"The fundamental change from quicksort to randomized quicksort is the addition of lines *R2a* and *R2b*. At line *R2a*, we randomly select an integer r between L and R, and at line *R2b* we swap $A[L]$ and $A[r]$."

"That's it? That's the only difference?" Tetra asked.

"That and the name, I suppose."

"If they're that similar, then the recurrence relation should be nearly unchanged too," I said.

> **Recurrence relation for RANDOMIZED-QUICKSORT**
>
> $$\begin{cases} \bar{T}_R(0) &= 4 \\ \bar{T}_R(1) &= 4 \\ \bar{T}_R(n) &= \dfrac{1}{n}\sum_{p=L}^{R}\left(7n + \underset{\sim}{4} + \bar{T}_R(p-L) + \bar{T}_R(R-p)\right) \end{cases}$$
>
> $$(n = 2, 3, 4, \ldots)$$

I continued. "If we use $j = 1, 2, \ldots, n-1, n$ in place of $p = L, L+1, \ldots, R-1, R$, we can rewrite the recurrence relation like this."

$$\begin{cases} \bar{T}_R(0) &= 4 \\ \bar{T}_R(1) &= 4 \\ \bar{T}_R(n) &= 7n + 4 + \dfrac{1}{n}\sum_{j=1}^{n}\left(\bar{T}_R(j-1) + \bar{T}_R(n-j)\right) \end{cases} \quad (n = 2, 3, 4, \ldots)$$

"So it looks like we're still where we were."

$$\bar{T}_R(n) = O(n \log n)$$

"Indeed," Miruka agreed, "but this has a different meaning. The $\bar{T}_R(n)$ here is *not* an average for an input with a uniform distribution. Randomly selecting the pivot means the number of execution steps doesn't rely on the input. In fact, with a randomized quicksort we can get different numbers of steps for the same input each time we run the algorithm. So $\bar{T}_R(n)$ isn't an average, it's an expected value. For any input of size n, we have an expected value for the number of execution steps for $\bar{T}_R(n)$, and it is at most of order $n \log n$."

∎ $\bar{T}_Q(n) = O(n \log n)$

> For quicksort, the *average number* of execution steps for <u>input with a uniform distribution</u> is at most of order $n \log n$.

- $\bar{T}_R(n) = O(n \log n)$

> For randomized quicksort, the *expected value* for the number of execution steps <u>for any input</u> is at most of order $n \log n$.

Tetra smiled. "That's so neat! Just randomly selecting the pivot frees us from the preconditions!"

10.4.3 Observing Comparisons

With her eyes closed, Miruka raised a finger and began tracing some strange shape in the air. We silently watched. After a time she finally opened her eyes and slowly spoke.

"We know that the quicksort algorithm sorts by repeatedly using a pivot to partition an array. It moves elements with values less than the pivot value to the left, and values that are at least the pivot value to the right."

"The left and right wings!" Tetra said.

"That's right, we create two wings. But maybe we still have something to learn about the element comparisons that are involved in that partitioning. Ready for a quiz?"

"I am!" Tetra said.

"Say you're using randomized quicksort to sort an array $\langle 1, 2, 3, \ldots, n \rangle$. Assuming $1 \leqslant j < k \leqslant n$, when are elements j and k compared?"

"Well, that depends, doesn't it?" Tetra said.

"It does indeed," Miruka said.

"I mean, they might be compared several times, or they might not be compared at all, right?"

"You tell me," Miruka said with a sly smile.

I thought on this while listening to them. *When will two given elements j and k be compared in a randomized quicksort?* Of course, they wouldn't be compared at all in some cases. For example, if we input $\langle 1, 2, 3 \rangle$ and randomly select 2 as the pivot, then during partitioning there would be comparisons of 1 and 2, and of 2 and 3, once each. When the partitioning was complete, the "left wing" would contain only 1, while the "right wing" would have only the 3 element. The 1 and the 3 would never be compared. Okay, so what about the general case, when the array was $\langle 1, 2, 3, \ldots, n \rangle$? When might j and k be compared then?

"Let's watch an example of partitioning," Miruka said. "The pivot is randomly selected, so the order of the elements within the wings isn't important. Let's treat the wings as sets of numbers. For example, say we're given an input of eight numbers."

$$\{1, 2, 3, 4, 5, 6, 7, 8\}$$

"If we select 5 as the pivot, the wings look like this."

$$\underbrace{\{1, 2, 3, 4\}}_{\text{left wing}} \quad \underbrace{5}_{\text{pivot}} \quad \underbrace{\{\ 6, 7, 8\ \}}_{\text{right wing}}$$

"What elements will have been compared to create these two wings? That's easy. Everything in the left wing is smaller than the pivot, and we know that because we compared each one with the pivot value of 5. All elements in the right wing are larger, and we know that the same way. In other words, in a single partitioning the comparisons are between the pivot and all other elements—no non-pivot elements are compared with each other."

"That's only for the first partitioning, though," I said.

"That's right. And once that's done, we start recursively sorting the left and right wings, neither of which contains the 5 we just used as a pivot. In other words, an element once selected as a pivot will never become one again."

- One of two compared elements is a pivot.

- Once an element has been selected as a pivot, it will not be selected again.

"In other words, we can see that in a single quicksort there will be at most one comparison between any two elements. So our elements j and k will be compared at most once, meaning 0 or 1 times. Okay, so let's take a closer look at these two wings we created."

$$\underbrace{\{1, 2, 3, 4\}}_{\text{left wing}} \quad \underbrace{5}_{\text{pivot}} \quad \underbrace{\{\ 6, 7, 8\ \}}_{\text{right wing}}$$

"Each of the wings will be recursively sorted, and when doing so there will be no comparisons between the $1, 2, 3, 4$ in the left wing and the $6, 7, 8$ in the right wing. There won't be a single between-wing comparison."

"It all makes so much sense when you describe it this way!" Tetra said.

I nodded. "Yeah, I never noticed any of this when playing with equations."

"So all of the comparisons take place inside of wings," Tetra said. "The new pivots are selected within wings, too, and the other elements inside are compared with that pivot. That's why there will never be a between-wing comparison."

"Exactly." Miruka stood. "Okay, now that we've taken a close look at how the comparisons take place, we can really get started. We can finally figure out when elements j and k will be compared in a quicksort."

"I know! When one of j or k is a pivot!" Tetra said.

"Can't be that," I said.

"No? Why not?"

"Just think about the example we just looked at. The pivot in the first partitioning was 5, and the second one was 3. Will 3 and 7 be compared? The 3 becomes the pivot in the second partitioning, but it will never be compared with the 7."

"Oh, of course! Because they got separated in the first partitioning!"

"So when will elements j and k be compared?" Miruka asked.

"I think she wants you to tighten up your conditions," I added.

"When one of j or k is a pivot, *and* when they don't get split into different wings?"

"Much better," I said. "Better still, in the range $j \leqslant \bigcirc \leqslant k$, elements j and k will be compared only when j or k are initially selected as the pivot."

"Wait, really? I'm not sure I see that..."

"Just think about what it means for j and k to get split up. If j and k were put into different wings, then there was some element p such that $j < p < k$, and furthermore p was selected as a pivot before either j or k. Conversely, j and k staying in the same wing would mean the opposite, that either j or k were selected as a pivot before any element p where j < p < k."

"Okay, now I get it."

"So the answer to Miruka's question, 'when are elements j and k compared?', is 'when either j or k are first selected as a pivot from among those elements from j to k.'"

Miruka gave a slight nod. "Correct. So now we have a better understanding of when j and k are compared. We can be particularly pleased with the fact that the number of comparisons will be either 0 or 1."

"Why's that?" Tetra asked.

"Because it should remind us of a good counting tool we have."

"An indicator," Lisa said.

"An indicator?" Tetra said.

"An indicator!" I said. "A variable that takes values of 0 or 1!"

"Specifically, an indicator random variable," Miruka continued. "We can expect a different number of j–k comparisons in each trial, but we know that it will be either 0 or 1. That's what makes it an indicator random variable. Let's call it $X_{j,k}$."

$$X_{j,k} = \begin{cases} 1 & \text{when elements j and k are compared} \\ 0 & \text{when elements j and k are not compared} \end{cases}$$

Tetra raised a hand. "Just to be sure, an indicator random variable is what we used when we were counting the number of heads in a coin flip, right?"

"That's right. They're a useful tool for counting numbers. Now we can use what we've learned to evaluate the expected value for the total number of comparisons, without solving the recurrence relation. That's possible because we can describe the number of comparisons as a sum. To do so, we use the fact that the expected value of a sum is the sum of expected values."

"Linearity of expectation," Lisa added.

10.4.4 Linearity of expectation

Miruka continued. "We're going to let X be a random variable describing the total number of element comparisons, and let $X_{j,k}$ be an indicator random variable describing the number of comparisons between elements j and k. Then X can be decomposed into the sum of $X_{j,k}$. Specifically, the sum of all j, k pairs where $1 \leqslant j < k \leqslant n$."

$$X = \sum_{j=1}^{n-1} \sum_{k=j+1}^{n} X_{j,k} \qquad \text{X as sum of } X_{j,k}$$

$$E[X] = E\left[\sum_{j=1}^{n-1} \sum_{k=j+1}^{n} X_{j,k}\right] \qquad \text{expected value of both sides}$$

$$= \sum_{j=1}^{n-1} E\left[\sum_{k=j+1}^{n} X_{j,k}\right] \qquad \text{from linearity of expectation}$$

$$= \sum_{j=1}^{n-1} \sum_{k=j+1}^{n} E[X_{j,k}] \qquad \text{from linearity of expectation}$$

"With this we have $E[X]$ as a sum for $E[X_{j,k}]$, so now we want to find the expected value for $X_{j,k}$. The fact that $X_{j,k}$ is an indicator random variable will be useful to that end, because the expected value of an indicator random variable is its probability."

10.4.5 The Expected Value of an Indicator Random Variable is its Probability

"What does that mean?" Tetra asked.

"Just look at the definition of an expected value," Miruka said.

$$E[X_{j,k}] = 0 \cdot \Pr(X_{j,k} = 0) + 1 \cdot \Pr(X_{j,k} = 1) \quad \text{def. of expected value}$$
$$= \Pr(X_{j,k} = 1) \qquad \qquad \text{0 terms disappear}$$

"In other words, the expected value $E[X_{j,k}]$ is just the probability that $X_{j,k} = 1$. Since $X_{j,k}$ is an indicator random variable showing whether elements j and k will be compared, we can say this."

$$E[X_{j,k}] = \text{probability of j and k comparison}$$

"Okay, so let's find that probability. Actually that's not so hard, since we already talked about *when* they're compared. Namely, when either j or k are first selected as a pivot from among those elements from j to k."

Probability of comparing j and k
= Whether either j or k are first selected as
a pivot from among those elements from j to k

"So what we want is the probability that j or k will be selected from among the $k-j+1$ elements from j to k."

$$\underbrace{j, j+1, \ldots, k-1, k}_{k-j+1 \text{ of these}}$$

"Think of this as a roulette wheel with $k-j+1$ slots, labeled with numbers $j, j+1, \ldots, k-1, k$. When we spin the wheel, what's the probability of the ball landing on j or k?"

"Should be this, right?" Tetra said.

$$\frac{\text{Ways of hitting j or k}}{\text{All possibilities}} = \frac{2}{k-j+1}$$

Miruka nodded. "That's right. Which gives us this."

$$E[X_{j,k}] = \text{Probability of comparing j and k} = \frac{2}{k-j+1}$$

"Okay then, back to the expected probability of a sum being the sum of expected probabilities."

$$\begin{aligned}
E[X] &= \sum_{j=1}^{n-1} \sum_{k=j+1}^{n} E[X_{j,k}] \\
&= \sum_{j=1}^{n-1} \sum_{k=j+1}^{n} \frac{2}{k-j+1} \qquad E[X_{j,k}] \text{ is a probability} \\
&= 2 \sum_{j=1}^{n-1} \sum_{k=j+1}^{n} \frac{1}{k-j+1}
\end{aligned}$$

"As k moves within the range $j+1 \leqslant k \leqslant n$, we'll have $k-j+1$ moving within the range $2 \leqslant k-j+1 \leqslant n-j+1$, so we can let

$m = k - j + 1$ to get this."

$$E[X] = 2\sum_{j=1}^{n-1}\sum_{m=2}^{n-j+1}\frac{1}{m}$$

$$= 2\sum_{j=1}^{n-1}\left(\sum_{m=1}^{n-j+1}\frac{1}{m} - \frac{1}{1}\right) \qquad \text{as a sum from 1}$$

$$= 2\sum_{j=1}^{n-1}(H_{n-j+1} - 1) \qquad \text{using } H_n$$

$$= 2\sum_{j=1}^{n-1}H_{n-j+1} - 2\sum_{j=1}^{n-1}1$$

$$= 2\sum_{j=1}^{n-1}H_{n-j+1} - 2(n-1)$$

"Similarly, when j moves within the range $1 \leqslant j \leqslant n-1$, we'll have $n-j+1$ moving within the range $2 \leqslant n-j+1 \leqslant n$, so we can let $\ell = n - j + 1$ to get this."

$$E[X] = 2\sum_{\ell=2}^{n}H_\ell - 2(n-1)$$

$$= 2(H_2 + H_3 + \cdots + H_n) - 2n + 2$$

$$\leqslant 2\underbrace{(H_n + H_n + \cdots + H_n)}_{n-1 \text{ of these}} - 2n + 2$$

$$= 2(n-1)H_n - 2n + 2$$

$$= 2n \cdot H_n - 2H_n - 2n + 2$$

$$= O(n \cdot H_n)$$

$$= O(n \log n)$$

"To review, we observed comparisons in randomized quicksort, then used linearity of expectation and the fact that the expected value of an indicator random variable is its probability to evaluate the order of the expected total number of comparisons. In the end, we found that we can expect a randomized quicksort to complete after an order of at most $n \log n$ comparisons."

$$E[X] = O(n \log n)$$

10.5 At the Restaurant

10.5.1 Other Randomized Algorithms

"We talked about RANDOM-WALK-3-SAT and RANDOMIZED-QUICKSORT," Tetra said, "but are there any other randomized algorithms?"

We were back at the restaurant near the station to grab some dinner.

"Infinitely many," Miruka replied. "The easiest to understand are randomized algorithms for understanding the overall state of things. If you have a dataset so large that it's hard to grasp its details, you can use random sampling to get at least some picture of what's going on with much less effort."

"Like tasting soup after giving it a good stirring!" Tetra said.

"There are also randomized algorithms for avoiding worst-case scenarios. Randomized quicksort is an example of that. As we saw, fixing the pivot location could result in a worst-case scenario, so we select it randomly instead."

"Interesting," I said.

"There are also randomized algorithms for obtaining some amount of proof. One example is probabilistic primality tests, algorithms for testing whether extremely large integers are primes. Their output is either 'definitely composite' or 'probably prime.'"

"It feels so strange to use 'probably' in math," Tetra said.

I agreed. "I think it's important that you clearly establish the probability of failure. You would want to say that the chance of being wrong is below some specific probability, rather than just shrug and say maybe."

"If you get an inconclusive results, you can also perform a more definitive test, though doing so will take time," Miruka said. "Nothing wrong with that. It's just a matter of balancing time versus precision."

"Trade-offs," Lisa muttered.

10.5.2 Preparation

"How's your conference presentation coming along?" I asked Tetra.

"Well, I was all set to talk about bubble sort and quicksort," she said, "but now that I've heard all this about randomized quicksort..."

"You're going to add another topic? You sure you'll have enough time for all that?"

"Ugh. Still no way I can get some more time?" Tetra asked, looking hopefully at Lisa beside her.

"Nope."

"Yeah, I figured."

"Handouts," Lisa said.

"Hey, that's a good idea," I said. "Just make some handouts that better describe anything you couldn't fully cover, detailed mathematical derivations and such, and give those to everyone who comes."

"Oh, that's great! Like a letter to everyone who came to hear my talk! I guess that does mean more time to get ready, though."

"I'll help," Lisa said.

"Glad to see you don't have stage fright, at least," I said.

"Oh, I do! But this time at least it will be a small audience, mostly junior high students, so I think I'll be okay."

"Have you reserved a room?" Miruka asked.

"Checking," Lisa said, opening her computer and tapping some keys.

"Wow, you can check from here?" Tetra said.

"Knowing Lisa, I'd bet she can check pretty much anything from here," Miruka said.

"Iodine," Lisa said.

"A lecture hall, huh?"

"A...lecture hall?!"

"Bigger audience," Lisa said.

10.6 AT THE NARABIKURA LIBRARY

10.6.1 Iodine

The weather forecast had called for clear skies on the day of the conference, but it was drizzling nonetheless. Of course. When Tetra, Yuri, and I stepped into the Narabikura Library we found Lisa there, waiting for us.

"Which way to the hall?" I asked.

"This way," Lisa said.

"Where's that super fine secretary?" Yuri asked.

Lisa pointed. "There." We looked in the direction she indicated and saw a tall man arranging books on a shelf. "Mr. Mizutani," Lisa added.

"Mizutani?" I said. "As in...the school librarian?"

"Brother," Lisa said.

"You're joking," I said, shocked.

"Wow, he's tall," Yuri said. "Speaking of which, where's Miruka?"

"I told you, she won't be here. She doesn't get back to Japan until tomorrow."

"Aww."

I looked at Tetra and realized she had been silent all morning.

"Getting nervous?" I asked.

"I'm fine," she said. "I've got my script all ready, so..."

Tetra showed me the bundle of papers she was clutching, but she didn't look fine at all.

10.6.2 Anxiety

The Iodine hall was full, mainly with junior and senior high school students.

"So many people!" Tetra said, scanning the room.

The room wasn't as large as the library's main hall, but the size clearly surprised Tetra, who likely had been imagining a more intimate classroom. Her session came after the morning lecture, a discussion of discrete mathematics presented by a university professor. During that first talk Tetra fidgeted next to me, scanning through her notes.

Eventually the first talk ended, and 'Randomized Quicksort' was displayed on the screen at the front of the room. It was Tetra's turn.

She left her seat and approached the stage, stumbling as she walked up the steps. Startled, I started to stand, but thankfully she didn't fall. She did, however, drop the stack of papers she had been carrying. She hurried to gather them and put them back in order. I felt so bad for her I couldn't bear to watch.

Eventually she collected herself, and bowed to the audience, who clapped.

"So, um, today I want to talk about the randomized quicksort algorithm, which, uh..."

And with that, she stopped.

She was clearly trying to speak, but no words came out. I could almost feel how nervous she was. The previously quiet audience began murmuring, everyone wondering what the problem was. Tetra remained frozen. I started to panic myself.

Yuri poked me. "Tetra needs help!" she whispered. As if I needed to be told.

"It's not like I can go up on the stage for her!" I hissed back.

From somewhere came a familiar citrus scent.

"Lisa, Tetra needs help!" said a familiar voice. I turned, and...

"Miruka?!"

"On it," Lisa said, opening her red computer. She tapped some keys, and after a time a loud tone, a sound like a tuning fork, could be heard throughout the hall. Lisa hit some more keys, and the screen behind Tetra changed.

> Quiet, please.

Everyone's eyes went to the screen. The buzz from the audience died down, and the room returned to silence. Lisa once again tapped on her keyboard, and the screen returned to how it was before.

"Tetra!" she called out in a husky voice. "Continue!"

10.6.3 The Presentation

Tetra snapped back. She took a deep breath. "Sorry about that. Today I want to talk about the randomized quicksort algorithm..." she continued, in a much more assured voice. I breathed a sigh of relief.

Having regathered, Tetra smoothly went through a description of terms, then an asymptotic analysis of the algorithm. She gave plenty of examples, limited her use of equations where she could, and made it through her talk at a brisk pace. The students in the audience, initially unsure, were clearly drawn in by her discussion and some key phrases she inserted here and there: "Examples are key to understanding." "It's best to start from the obvious." "Generalization through introduction of a variable." "Quantitative evaluation with clear preconditions." She even gave a demonstration of the algorithm in action.

Before long, she was actually smiling.

When she said, "From this, we can see that there will never be a comparison between the two wings," I saw audience members nodding, and heard one "Interesting." Several people near me were taking careful notes.

Tetra's anxiety dissipated and her eyes took on a twinkle as she continued through a series of key concepts. Linearity of expectation. The expected value of a sum is a sum of expected values. The expected value of an indicator random variable is a probability. Indicator random variables are a counting tool.

Everyone in the hall seemed to be having as much fun as she was.

10.6.4 Passing on

"And that concludes my presentation of the randomized quicksort algorithm. But, if I may, I would like to close with some comments about things I've been feeling lately.

"Mathematics isn't about solving problems that someone has given you. It's about wanting to find solutions to things you don't understand, things that are complicated, things that are unclear. That's an important feeling to have.

"You need to set up your own problems, work on them by yourself, think about them for yourself. When you do so, you sometimes make wonderful discoveries. Sorting algorithms are all about putting things in order, so I never dreamed that randomization could make that easier. I mean, isn't that just amazing? That feeling of amazement is what I hope you'll leave with today.

"I had a lot of help getting ready for this presentation today. From mathematicians who died long ago and those who did their work fairly recently, from teachers at my school, and from my schoolmates, my friend Lisa in particular. I'd like to take this opportunity to thank all of them.

"Please excuse me for having gone on for so long. I will close with something I learned from a classmate who I truly lo—uh, truly respect. There was a mathematician in the seventeenth century named Leibniz who studied binary numbers. Of course he knew nothing then about twenty-first century computers, but this research by him and many other mathematicians over the ages is present in the computers we use today. Leibniz the mathematician has passed, but his mathematics lives on, his gift from so long ago.

"Mathematics transcends time.

"Math, the math that we produce, extends into the future. Through my presentation today, maybe I passed on some tiny bit of something to you. I truly hope that you will all go out and pass on some of the math that you have learned to others. Pass on the wonder of mathematics, the wonder of learning, the wonder of experiencing wonder.

"So I will leave you with those words: Mathematics transcends time.

"Thank you."

Tetra bowed to the audience, who responded with hearty applause.

"Oh, wait! Wait!" she shouted, flapping her hands to call for silence. "I forgot to teach you the Fibonacci sign!" After a quick explanation of the math-lover's handsign, she flipped her fingers "1...1...2...3..." The audience responded with a gleeful "5," and then another round of applause.

And thus Tetra's presentation ended on a very high note.

10.6.5 Oxygen

"I can't believe I did that!" Tetra said. "It's like my brains just switched off!"

"More practice," Lisa said, reaching for her glass of milky iced tea.

We had gathered for lunch in Oxygen, the cafe in the Narabikura Library. The rain had finally ceased, and the sky even showed a few patches of blue.

"You recovered, that's what's important," Miruka said.

"You even taught them the Fibonacci sign," I said.

"I was *sooo* nervous," Tetra said, scanning the the conference pamphlet. "Honestly I just want to spend the rest of the day relaxing, but then I'd miss these other interesting sessions."

"There's even some others for junior high students," Yuri said. I noticed she was nervously looking around the room.

Several people we didn't know came up to our table as we ate our lunch. Each was holding the handout from Tetra's presentation. These included a boy from China and a girl from Sweden, showing how math transcends not only time but also national boundaries. The girl spoke in English to Tetra, who replied in kind.

"Wow, Tetra, that was amazing," I said after she left.

"I'm okay with the English, but the math she was talking about... Miruka, what's the difference between 'probabilistic analysis of algorithms' and 'analysis of randomized algorithms'?"

"The first is about the execution time for an algorithm assuming some probability distribution for the input. The second doesn't require any assumptions about the input, like the randomized quicksort algorithm."

10.6.6 Connecting

"By the way, Miruka," Tetra said, "the other day you were talking about academic papers as a way of passing things on to the future, but I think there are other ways. Teaching, for one. When you teach someone something, they can go on and teach other people. And so on and so on."

Miruka folded her arms. "Hmph."

"Yeah," Yuri sighed.

"What's up with you, Yuri? You seem kinda down," I said.

"Just wondering how I could pass on my own message," she said, slumping into her seat. "Especially to someone far away..."

"You can do it," Tetra said. "Distance is no problem. You just need the right words."

Just then someone called Yuri's name. She looked up with wide eyes and scanned the room until she saw a boy at another table, apparently another junior high school student, waving at her.

"Why didn't you tell me you'd come?" she half shouted, leaving her seat and trotting to his table.

"Her first kiss, you know," Tetra whispered.
"Her first... what?!"
"If foreheads count."

10.6.7 In the Garden

After we finished our lunch, Miruka and I went for a walk in the garden. The air was still damp from the recent rain.

"You don't have to play daddy to Yuri, you know," Miruka said.

"Never mind that," I said. "What are you doing back?"

She took my hand and squeezed it tight. "I came back a day early."

I wasn't sure how to respond.

"Cold, aren't you."

"Sorry. I, uh..." *Still don't know what to say.* "I'm glad."

"I meant your hand."

"Oh, right." *Gotta say more.* "Listen, Miruka. I don't know what I can promise you. But..." My head was spinning. I wasn't even sure what I was saying.

Miruka sighed. "You don't understand anything."

"I— Huh?"

"To me, you are—" She stopped.

"I'm what?"

Miruka looked away. "Never mind."

"How about this. I promise that someday I'll make you a promise."

"A meta-promise, huh? Well, then make me a pre-promise too."

"A pre-promise?"

Miruka took a step closer to me. "No telephone calls." I reflexively took a step back. "And no letters. Tell me, what does it mean to be close to one another?"

"What does it mean...?"

Miruka pulled on my hand and again stepped closer. "You don't know?"

"I..."

"You will soon." She brought her face closer to mine. I noticed that her metal framed glasses had slightly blue-tinted lenses. I felt lost in the eyes behind them. "In less than one minute."

10.6.8 Symbol of a Promise

Two minutes later Tetra entered the garden. "There you two are! Oh, look, a rainbow!"

The skies had completely cleared, and indeed a large rainbow had appeared.

"Come on guys, the afternoon session's about to start!"

As we headed back inside I turned back for another look at the rainbow stretching across the blue sky. It did indeed seem to portent a great promise. The rainy season would soon be over, and the long, hot summer would begin.

> It has been some thirty years since the concept of randomized algorithms was born, and today they have obtained full citizenship in the field of algorithmic information theory. Indeed, they are no longer differentiated from "normal" algorithms. Even so, it is difficult to claim that in the world of practical algorithms they have received sufficient recognition of their effect and value.
>
> HISAO TAMAKI
> *Randomized Algorithms* [24]

Epilogue

"Hey!" she shouted, bounding into the teachers's office.

"What's up?" he said.

"Today I'm the one giving problems!"

> Kiosks A and B both sell lottery tickets. What can you say about a rumor that kiosk A sells more winning lottery tickets than kiosk B does?

"Mathematical nonsense," he immediately said. He wasn't one to be tricked so easily.

"Except that this rumor is true!" she said with a smirk.

"Impossible. Unless someone is cheating."

"Nope. It's true with no cheating. In fact, kiosk A just sells more tickets! The more a kiosk sells, the better the probability of it selling a winning ticket. Gotcha!"

"But the probability that any given ticket wins hasn't changed at all."

"Doesn't matter. The rumor morphed and more people started buying tickets from kiosk B, so in the end both ended up the same."

The teacher sighed.

"Oh, come on," she teased. "Not every problem has to be so serious. Speaking of which, got a new card for me?"

"Hmm, try this one next."

"What's this? 'If you replace a single value in a random number table, is it still a random number table?'"

"Do you know what a random number table is?"

```
8 0 0 5 8 9 6 7 7 0 2 9 7 5 9 6 8 5 1 4
5 8 2 7 7 2 1 7 6 6 0 8 1 5 6 2 2 3 6 1
5 2 8 9 9 2 0 7 5 0 1 0 1 6 8 9 8 9 6 7
3 5 1 9 4 6 2 9 8 9 7 7 1 1 3 6 3 9 2 2
9 4 8 6 5 8 4 7 5 4 5 1 5 7 9 4 4 1 9 9
4 0 4 9 7 3 5 0 1 3 8 2 6 2 0 3 8 7 7 5
3 5 6 3 1 3 4 8 7 2 2 0 3 8 5 5 1 8 4 8
2 9 3 8 4 5 9 0 7 6 0 2 9 5 4 6 0 6 4 0
1 8 7 0 5 6 1 4 7 2 6 6 1 5 9 3 1 8 0 2
5 8 7 1 0 3 5 8 4 6 6 1 6 1 9 5 6 7 ···
```

"Just what it says, right? A series of rows and columns filled with randomly generated numbers. This one looks big enough that just one value wouldn't make much of a difference."

"So you're saying that a random number table is still a random number table after you've replaced one value?"

"Sure, I guess."

"So what if you take the new table and replace one more value? Is it still a random number table then?"

Her expression became less confident. "Hmm..."

"Aren't you in essence saying you can replace an arbitrary number of values, and still call it a random number table?"

"Well when you put it that way..."

"Take your time and think about it at home. But it looks like the rain finally let up, so for now you should enjoy the outdoors."

"Yeah. Random numbers aside," she said, playing with the card, "tell me more about this 'examples are key to understanding' thing you're always saying."

"In retrospect, maybe I should have said they're a touchstone for understanding."

"What's a touchstone?"

"A kind of stone people used to use to assay precious metals."

"Like, a way to test if you've struck gold?"

"Exactly. The idea is that if you can create an example of a thing, you must understand it."

"I wonder if there's a test like that I can apply to myself?"

"In what sense?"

"A test to make sure my future is golden. College entrance exams don't exactly feel like a touchstone for success in life."

"Indeed. College entrance exams are just a selection mechanism for universities, nothing more."

"I just... I dunno. I'm not super confident about how things will go from there."

Her expression took on an even more serious cast. As a teacher he felt he should be able to give her a better answer, but he was also aware of the limitations of his position. The problems one faced at seventeen years old demanded self-exploration of their solutions.

"I know how you feel," he finally said. "I felt the same way at your age."

"Seriously?"

"You bet."

"So do things get easier when you get old?"

"Hey now, watch who you're calling old."

"You know what I mean. It's just, I feel like I've been allotted only so much time for my life, but I'm not sure what I should do with it. I like math, and I guess I'm a pretty good student, but it's hard to figure out what happens after college entrance exams. And nobody will tell me! Even worse, the more I read, the more I think, the more I study... the more I realize how little I know. So what's it all about? Get into a good school so I can get into a good company and get a good job and meet a good guy? Is that what's supposed to make me happy?"

"That's for you to say."

"Are you glad you became a math teacher?"

"Sure. I love teaching, I love math, I love talking with my students. I enjoy serious problems with serious solutions. Over centuries, millennia even, humanity has spent countless hours studying and teaching math, and through this job I've gotten a glimpse of why we have."

"Huh."

"Also, math transcends time."

"So it allows you to touch infinity?"

"Touch infinity, I like that. Touching infinity through the present."

"Right! Because the present is where we always are. Thanks for that, I'm feeling much better now."

"Glad to be of service."

"I'm outta here. See you soon!"

"Later."

The girl started to leave the office, but turned at the doorway. "Hey, I've thought of another answer!"

"To what?"

"The random number table problem."

"Ah. So what's your answer?"

"That there's no such thing as a random number table!"

"A bold statement."

She grinned. "Alternatively, that there's a problem with assuming that tables either are or aren't random. Maybe the better approach is to consider degrees of randomness, not a binary state."

"Very interesting."

"Okay, enough for today. Now I'm really off. Thanks!"

She flicked her fingers and headed out.

The teacher watched out the window, and eventually saw her passing through the school gates. She turned and gave an exaggerated wave, which he returned. *Such a bright child*, he thought, and felt arise within him a prayer for her success, that she would realize her dreams.

He looked up at the post-rain sky, deepest blue and adorned with a promising rainbow.

The last decade has witnessed a tremendous growth in the area of randomized algorithms. During this period, randomized algorithms went from being a tool in computational number theory to finding widespread application in many types of algorithms. Two benefits of randomization have spearheaded this growth: simplicity and speed. For many applications, a randomized algorithm is the simplest algorithms available, or the fastest, or both.

PRABHAKAR RAGHAVAN AND RAJEEV MOTWANI
Randomized Algorithms [25]

Afterword

> One's writing never includes everything one intended to include.
>
> AYA KODA
> *Tsutsumu*

Thank you for reading *Math Girls 4: Randomized Algorithms*. This is the fourth book in the *Math Girls* series, preceded by *Math Girls,* which I wrote in 2007, *Math Girls 2: Fermat's Last Theorem* (2008), and *Math Girls 3: Gödel's Incompleteness Theorems* (2009).

Like its predecessors, this book followed the activities of its narrator and math girls Miruka, Tetra, and Yuri, along with a new member in their cadre: the husky-voiced, reticent Lisa. As usual, we watched as they tackled problems in mathematics and in youth, activities that I did my best to capture as I wrote. Each takes their own approach to solving the problems they face, sometimes arriving at an answer, sometimes stumbling along the way. While writing, every day brought startling new discoveries and the thrill of not knowing what would happen next. The entire time, I found myself being surprised at some things they said and did, scolding them for others. I hope that my readers have experienced something similar.

As with previous books in this series, this book was created using LaTeX 2_ε and the AMS Euler font. Also as before, Haruhiko Okumura's book *Introduction to Creating Beautiful Documents with*

LaTeX 2_ε was an invaluable aid during layout, and I thank him for it. I also thank Kazuhiro Okuma (a.k.a. tDB) for his elementary mathematics handout macro, emath, which I used to create the diagrams. I also thank the following persons for proofreading and giving me invaluable feedback (though it goes without saying that any remaining errors are the responsibility of the author[1]).

> actuary_math, Ryo Akazawa, Tetsuya Ishiu, Kazuhiro Inaba, Ryuhei Uehara, Hiromichi Kagami, Toshiki Kawashima, Kazuhiro Kezuka, Kayo Kotaki, Haruaki Tasaki, Hiroaki Hanada, Yoichi Hirai, Hiroshi Fujita, Masahide Maehara, Tadanori Matsuki, Kiyoshi Miyake, Kenta Murata (mrkn), Kenji Yamaguchi, Tsutomu Yano, Yuko Yoshida

I thank my readers and the visitors to my website, and my friends for their constant prayers.

I thank my editor, Kimio Nozawa, for his continued support during the long process of creating this book.

I thank the many readers of the *Math Girls* series for their support and comments; your encouragement is a treasure beyond description.

I thank my dearest wife and my two sons.

I dedicate this book to my father, from whom I inherited a love of learning.

Finally, thank *you* for reading my book. I hope that someday, somewhere, our paths shall cross again.

Hiroshi Yuki
2011
http://www.hyuki.com/girl/

---

[1] Translator's note: Or the translator!

Recommended Reading

> "The library was like a second home. Or maybe more like a real home, more than the place I lived in."
>
> HARUKI MURAKAMI
> *Kafka on the Shore*

[Note: The following references include all items that were listed in the original Japanese version of *Math Girls 4: Randomized Algorithms*. Most of those references were to Japanese sources. Where an English version of a reference exists, it is included in the entry.]

GENERAL READING

[1] Yuki, H. (2011). *Math Girls*. Bento Books. Published in Japan as *Sūgaku gāru* (Softbank Creative, 2007).

> The story of two girls and a boy who meet in high school and work together after school on mathematics unlike anything they find in class.

[2] Yuki, H. (2012). *Math Girls 2: Fermat's Last Theorem*. Bento Books. Published in Japan as *Sūgaku gāru / Ferumā no saishū teiri* (Softbank Creative, 2008).

> The second book in the *Math Girls* series, where new "math girl" Yuri joins the others on a quest to understand "the true form" of the integers. Presents groups, rings, and fields among other topics, building to a tour of Fermat's last theorem.

[3] Yuki, H. (2016). *Math Girls 3: Gödel's Incompleteness Theorems*. Bento Books. Published in Japan as *Sūgaku gāru / Gēderu no saishūteiri* (Softbank Creative, 2009).

> The third book in the *Math Girls* series, where the math girls (and boy) use formal systems to learn "the mathematics of mathematics" and about Gödel's oft-misunderstood theorems.

[4] Yuki, H. (2005). *Purograma no sūgaku* [Math for Programmers]. Softbank Creative.

> A book that teaches mathematical approaches that are useful for computer programming. Includes descriptions of logic, mathematical induction, permutations and combinations, and induction.

[5] Ito, K. (2010). *Kakuritsuron to watashi* [Probability Theory and Me]. Iwanami Shoten.

> A collection of essays by the mathematician Kiyoshi Ito, a pioneer in stochastic differential equations. Also contains descriptions of Kolmogorov's theories of education and the attitude one should take when learning mathematics.

DISCRETE MATHEMATICS

[6] Matoušek, J. and Nešetřil J. (2008). *An Invitation to Discrete Mathematics, Vol. 1* (2nd Ed.). Translated by Negami, S. and

Nakamoto, A. as *Risan sūgaku e no shōtai.* (1st Ed., Springer Japan, 2002).

> A good reference book for learning the fundamental techniques of asymptotic approximation, basic quantitative evaluation, and combinations. (I referenced this book regarding the evaluation of $n!$ in chapter 6.)

[7] Matoušek, J. and Nešetřil J. (2008). *An Invitation to Discrete Mathematics, Vol. 2* (2nd Ed.). Translated by Negami, S. and Nakamoto, A. as *Risan sūgaku e no shōtai.* (1st Ed., Springer Japan, 2002).

> The second volume in this series, teaching graph theory, projective planes, probability, and generating functions. (I referenced this book when writing chapters 4 and 5.)

[8] Graham, R., Knuth, D., and Patashnik, O. (1994). *Concrete Mathematics: A Foundation for Computer Science (2nd Edition).* Addison-Wesley Professional. Translated by Arisawa, M., Yasumura, M., Akino, T., and Ishihata, K. as *Konpyūtā no sūgaku* (Kyoritsu Shuppan, 1993).

> A textbook on discrete mathematics, with finding sums as its theme. (I referenced this book when writing about discrete probability in chapter 8 and asymptotic approximation in chapter 9.)

PROBABILITY THEORY

[9] Kolmogorov, A., Zhurbenko I., and Prokhorov A. (1995). *Vvedenie v teoriyu veroiatnostei* [Introduction to Probability Theory] (in Russian). Translated by Maruyama, T. and Baba, Y. as *Korumogorofu no kakuritsu-ron nyūmon.* (Morikita, 2003).

> An introduction to probability theory written by Kolmogorov, the mathematician who developed axiomatic probability theory. Allows you to experience the vibrance of probability through many examples and easy problems. (I referenced this book when writing chapters 4, 5, and 8.)

[10] Hiraoka, K. and Hori, G. (2009). *Puroguramingu no tame no kakuritsu tōkei* [Probability and Statistics for Programming]. Ohmsha.

> A reference book aimed at mathematical laypersons for learning the basics of probability and statistics. Teaches the basis for application of these tools, starting from learning how "probability is an area." (I referenced this book while writing chapters 4 and 5).

[11] Blom, G., Holst, L., and Sandell, D. (1994). *Problems and Snapshots from the World of Probability*. Translated by Mori, M. as *Kakuritsu-ron e yōkoso* (Springer Japan, 2005).

> A collection of problems for grasping an overall picture of probability theory through solving typical problems. (I referenced this book while writing chapters 4, 5, and 8).

[12] Koharu, A. (1973). *Kakuritsu tōkei nyūmon* [Introduction to Probability and Statistics]. Iwanami Shoten.

> A textbook for probability and statistics. (I referenced this book while writing chapters 4, 5, and 8).

LINEAR ALGEBRA

[13] Tahakashi, M. (1989). *Gyōretsu (Monogurafu 8)* [Matrices (Mathematics Monographs Vol. 8)]. Kagaku Shinko-shinsha.

> A good reference and problem book for high school students to learn the basics of matrices. (I referenced this book while writing chapter 7.)

[14] Hiraoka, K. and Hori, G. (2004). *Puroguramingu no tame no senkei daisū* [Linear Algebra for Programming]. Ohmsha.

> A reference book aimed at mathematical laypersons for practical applications of linear algebra. (I referenced this book while writing chapter 7.)

[15] Shiga, K. (1988). *Senkei daisū sanjū-kō* [Thirty Lectures on Linear Algebra]. Asakura Shoten.

> A reference book that teaches linear algebra in thirty steps, from a systems of equations to eigenvalues. Written with the goal having readers "step up onto the stage of linearity" so that they can grasp a broad view of linear algebra. (I referenced this book while writing chapter 7.)

[16] Shiga, K. (2009). *Senkei to iu kōzō e: Jigen o koete* [The Structure of Linearity: Beyond Dimensions]. Kinokuniya.

> A reference book for learning about linearity in a two-part composition: "finite dimensional linearity" and "infinite dimensional linearity." (I referenced this book while writing chapter 7.)

ALGORITHMS

[17] Bentley, J. (1999). *Programming Pearls*. Translated by Kobayashi, K. as *Shugyoku no puroguramingu* (Pearson Education, 2000).

> A collection of very interesting topics related to algorithms. (I referenced this book while writing chapter 10.)

[18] Cormen, T., Leiserson, E., Rivest, R., and Stein, C. (2009). *Introduction to Algorithms, 3rd Ed.* MIT Press.

> A standard textbook on algorithms. (I referenced this book when writing chapters 2, 6, and 10.)

[19] Knuth, D. (1997). *The Art of Computer Programming Volume 1: Fundamental Algorithms, 3rd Ed.* Translated by Arisawa, M. and Wada, E. under the same title (ASCII, 2004).

> A historically important text sometimes called "the Bible of algorithms." This first volume describes discrete mathematics and data structures. (I referenced this book when writing chapter 2.)

[20] Knuth, D. (1997). *The Art of Computer Programming Volume 2: Seminumerical Algorithms, 3rd Ed.* Translated by Arisawa, M. and Wada, E. under the same title (ASCII, 2004).

> The second volume of "the Bible of algorithms" discusses random numbers and arithmetic operations. (I referenced this book when writing chapters 9 and 10, and practice problem 44 in section 3.5 for the random table problem in the Epilogue.)

[21] Knuth, D. (1998). *The Art of Computer Programming Volume 3: Sorting and Searching, 2nd Ed.* Translated by Arisawa, M. and Wada, E. under the same title (ASCII, 2006).

> The third volume of "the Bible of algorithms" discusses sorting and searching. (I referenced this book when writing chapters 2, 6, and 10.)

[22] Knuth, D. (2008). *The Art of Computer Programming, Volume 4, Fascicle 0: Introduction to Combinatorial Algorithms and Boolean Functions.* Translated by Wada, E. under the same title (ASCII Mediaworks, 2009).

> Installment 0 of the fourth volume of "the Bible of algorithms." (The Strong–Correct–Beautiful–Kind problem in chapter 9 references the equation by Rivest in section 7.1.1 of this book.)

[23] Kleinberg, J. and Tardos, É. (2005). *Algorithm Design.* Translated by Asano, T., Asano, Y., and Hirada, T. as *Arugorizumu dezain* (Kyoritsu Shuppan, 2008).

> A reference book for learning typical ways of thinking about algorithms and how to design them. (I referenced this book when writing chapters 2, 6, and 10.)

RANDOMIZED ALGORITHMS

[24] Tamaki, H. (2008). *Rantaku arugorizumu* [Randomized Algorithms]. Kyoritsu Shuppan.

> The first text in Japan to take randomized algorithms as its primary focus. It classifies randomized algorithms into several patterns, and describes each in an easy-to-understand manner. (I referenced this book while writing chapters 9 and 10.)

[25] Motwani, R. and Raghavan, P. (1995). *Randomized Algorithms.* Cambridge University Press.

> A pioneering textbook on randomized algorithms. (I referenced this book when writing chapters 9 and 10.)

[26] Mitzenmacher, M. and Upfal, E. (2005). *Probability and Computing: Randomized Algorithms and Probabilistic Analysis*. Translated by Koshiba, T. and Kawachi, A. as *Kakuritsu to keisan: Rantaku arugorizumu to kakuritsu-teki kaiseki* (Kyoritsu Shuppan, 2009).

> A textbook on analysis of randomized algorithms and stochastic analysis of algorithms. Covers an extensive array of topics, from the fundamentals of probability theory to advanced topics in randomized algorithms. (I referenced this book when writing chapters 8, 9, and 10.)

[27] Hromkovič, J. (2004). *Algorithmics for Hard Problems*. Translated by Wada, K., Masuzawa T., and Motoki, M. as *Keisan kon'nan mondai ni taisuru arugorizumu riron* (Springer Fairlark Tokyo, 2005).

> A reference book that describes deterministic, approximate, and probabilistic approaches to designing algorithms to address difficult problems. (I referenced this book when writing chapter 10.)

ACADEMIC PAPERS

[28] Schöning, U. (1999). A probabilistic algorithm for k-SAT and constraint satisfaction problems. *Proceedings of the 40th Annual IEEE Symposium on Foundations of Computer Science*. IEEE Computer Society.

> A paper describing k-SAT, a generalization of the 3-SAT problem. It presents a method for probabilistically solving the k-SAT problem by using a random walk, and furthermore describes it as a generalized constraint satisfaction problem (CSP). (I referenced this paper

when writing chapter 9. Note that the evaluation in chapter 9 also references Ash, R. (1990) *Information Theory*, Dover.)

[29] Dawkins, B. (1991). Siobhan's problem: The coupon collector revisited. *The American Statistician*, vol. 45, No. 1.

> A paper that addresses the coupon collector problem. (I referenced this paper for problem 5-3, "Expected value until all faces come up.")

WEBSITES

[30] Yuki, H.: *Math Girls* http://www.hyuki.com/girl/en.html.

> The English version of the author's *Math Girls* web site.

"So no, I'm not alone, even if we're working on different problems. By thinking through things by myself, by fighting through the rough spots on my own, I grow closer to you."

HIROSHI YUKI
Math Girls 4: Randomized Algorithms

Index

A
algorithm, 24
asymptotic analysis, 48, 167

B
BINARY-SEARCH, 181
binomial distribution, 142
binomial theorem, 74
BUBBLE-SORT, 191

C
calculation model, 38
Cayley–Hamilton theorem, 276
certain event, 111
characteristic equation, 290
clause, 307
combination, 70, 303
comparative search, 206
comparative sort, 200
comparison tree, 201
conjunctive normal form (CNF), 308

coupon collector's problem, 159

D
determinant, 235
diagonal matrix, 284
diagonalization, 284
DICE-GAME, 97
disjoint set, 113

E
eigenvalue, 291
eigenvector, 291
events, 108
exclusivity (of variables), 333
execute, 27
expansions, 76
expected value, 123, 130, 382

H
Hamming distance, 320

I
inconsistent, 213

independence (of variables), 333
indicator, 42, 185
indicator random variable, 145, 386
inner product, 232
intersection (of sets), 112
inverse (of a matrix), 234, 285

K
Kirchoff's law, 195

L
linear search, 25
linearity of expectation, 136, 386
LINEAR-SEARCH, 27
literal, 307
logical expression, 308

M
mapping, 40
Monte Carlo algorithm, 324
Monty Hall problem, 14
mutually exclusive events, 113

N
normalization, 115
NP-complete problem, 311
NP problem, 310

P
P versus NP, 310
permutation, 57, 58
P problem, 310
the piano problem, 257
probability, 106, 108
probability axioms, 108

probability distribution, 106, 108
procedure, 28
pseudocode, 27

Q
quantitative evaluation, 47
QUICKSORT, 351

R
random variable, 123, 124
random walk, 273
RANDOMIZED-QUICKSORT, 381
RANDOM-WALK-3-SAT, 315
recursion, 357
regular, 215
ROULETTE-GAME, 99

S
sample space, 105, 108
satisfiability problem, 305
satisfied, 309
sentinel, 45
SENTINEL-LINEAR-SEARCH, 44
Stirling's approximation, 337

T
trace, 278
tree diagram, 58
trial, 125

U
underdetermined, 215
union (of sets), 113

V
variable, 306

W
walkthrough, 28
Wandering Alice problem, 279

Other works by Hiroshi Yuki

(in English)

- *Math Girls*, Bento Books, 2011
- *Math Girls 2: Fermat's Last Theorem*, Bento Books, 2012
- *Math Girls 3: Gödel's Incompleteness Theorems*, Bento Books, 2016
- *Math Girls Manga, Vol. 1*, Bento Books, 2013
- *Math Girls Manga, Vol. 2*, Bento Books, 2016
- *Math Girls Talk About Equations & Graphs*, Bento Books, 2014
- *Math Girls Talk About the Integers*, Bento Books, 2014
- *Math Girls Talk About Trigonometry*, Bento Books, 2014

(in Japanese)

- *The Essence of C Programming*, Softbank, 1993 (revised 1996)
- *C Programming Lessons, Introduction*, Softbank, 1994 (Second edition, 1998)

- *C Programming Lessons, Grammar*, Softbank, 1995
- *An Introduction to CGI with Perl, Basics*, Softbank Publishing, 1998
- *An Introduction to CGI with Perl, Applications*, Softbank Publishing, 1998
- *Java Programming Lessons (Vols. I & II)*, Softbank Publishing, 1999 (revised 2003)
- *Perl Programming Lessons, Basics*, Softbank Publishing, 2001
- *Learning Design Patterns with Java*, Softbank Publishing, 2001 (revised and expanded, 2004)
- *Learning Design Patterns with Java, Multithreading Edition*, Softbank Publishing, 2002
- *Hiroshi Yuki's Perl Quizzes*, Softbank Publishing, 2002
- *Introduction to Cryptography Technology*, Softbank Publishing, 2003
- *Hiroshi Yuki's Introduction to Wikis*, Impress, 2004
- *Math for Programmers*, Softbank Publishing, 2005
- *Java Programming Lessons, Revised and Expanded (Vols. I & II)*, Softbank Creative, 2005
- *Learning Design Patterns with Java, Multithreading Edition, Revised Second Edition*, Softbank Creative, 2006
- *Revised C Programming Lessons, Introduction*, Softbank Creative, 2006
- *Revised C Programming Lessons, Grammar*, Softbank Creative, 2006
- *Revised Perl Programming Lessons, Basics*, Softbank Creative, 2006

- *Introduction to Refactoring with Java*, Softbank Creative, 2007

- *Math Girls / Fermat's Last Theorem*, Softbank Creative, 2008

- *Revised Introduction to Cryptography Technology*, Softbank Creative, 2008

- *Math Girls Comic (Vols. I & II)*, Media Factory, 2009

- *Math Girls / Gödel's Incompleteness Theorems*, Softbank Creative, 2009

- *Math Girls / Randomized Algorithms*, Softbank Creative, 2011

- *Math Girls / Galois Theory*, Softbank Creative, 2012

- *Java Programming Lessons, Third Edition (Vols. I & II)*, Softbank Creative, 2012

- *Etiquette in Writing Mathematical Statements: Fundamentals*, Chikuma Shobo, 2013

- *Math Girls Secret Notebook / Equations & Graphs*, Softbank Creative, 2013

- *Math Girls Secret Notebook / Let's Play with the Integers*, Softbank Creative, 2013

- *The Birth of Math Girls*, Softbank Creative, 2013

- *Math Girls Secret Notebook / Round Trigonometric Functions*, Softbank Creative, 2014

- *Mathematical Writing, Refinement Edition*, Chikuma Shobo, 2014

- *Math Girls Secret Notebook / Chasing Derivatives*, Softbank Creative, 2015

www.ingramcontent.com/pod-product-compliance
Ingram Content Group UK Ltd.
Pitfield, Milton Keynes, MK11 3LW, UK
UKHW041004030325
4827UKWH00015B/130